Praise for *Laughter: A Scientific Investigation*

"Life without laughter would not be worth living, yet science is just beginning to study this striking human trait. Robert Provine, the world's expert on the psychology and biology of laughter, has written a lively and fascinating introduction filled with provocative insights and surprising discoveries . . . written with the warmth and light touch that the subject deserves."
—Steven Pinker, author of *How the Mind Works*

"A beautifully written book . . . written with panache and humor."
—V. S. Ramachandran, co-author of *Phantoms of the Brain*

"Provine's well-written, often amusing and always fascinating exposé presents laughter in all its complexity and with all its contradictions."
—*Scientific American*

"Readers may laugh out loud. . . . Full of questions that tickle a reader's curiosity."
—*The Dallas Morning News*

"*Laughter* is no joking matter. Finally, a serious book about the funny side of life. Facts, theories, and even some tips about better living through laughter. A joy to read."
—Joseph LeDoux, author of *The Emotional Brain*

"[Provine's] fascinating book covers everything from the evolution of homo sapiens' laugh-producing vocal-cords to the engineering of the first laugh tracks in the 1950s. . . . This is one of those rare books that actually changes a reader's perception of his surroundings . . . and a reminder to readers that there's always a little more to notice in day-to-day social exchange than we ordinarily let ourselves see."
—*Seattle Union Record*

"It is hard to overstate how overdue Provine's book is, or how good it is. . . . A masterful job of collecting fascinating anecdotes. . . . A joy to read."
—James W. Kalat, author of *Biological Psychology* (seventh edition)

"A groundbreaking, fun-to-read anthropological study of laughter . . . fresh and revealing."
—*Booklist*

"Crisply written and often hilarious."
—*Evening Standard* (London)

"Robert Provine's investigation of laughter . . . is a model of constructive scientific thinking."
—*The Times* (London)

PENGUIN BOOKS

LAUGHTER

Robert R. Provine has authored more than fifty research articles concerning developmental neuroscience and animal and human behavior. His findings on laughter have been featured in dozens of articles worldwide, appearing in *The New York Times, Newsweek, The Daily Telegraph, New Scientist, The Observer, Discover*, and the *Los Angeles Times.*

ROBERT R. PROVINE

PENGUIN BOOKS

PENGUIN BOOKS
Published by the Penguin Group
Penguin Putnam Inc., 375 Hudson Street,
New York, New York 10014, U.S.A.
Penguin Books Ltd, 80 Strand,
London WC2R 0RL, England
Penguin Books Australia Ltd, Ringwood,
Victoria, Australia
Penguin Books Canada Ltd, 10 Alcorn Avenue,
Toronto, Ontario, Canada M4V 3B2
Penguin Books (N.Z.) Ltd, 182–190 Wairau Road,
Auckland 10, New Zealand

Penguin Books Ltd, Registered Offices:
Harmondsworth, Middlesex, England

First published in the United States of America by Viking Penguin,
a member of Penguin Putnam Inc., 2000
Published in Penguin Books 2001

1 3 5 7 9 10 8 6 4 2

Illustrations on pages 78, 82, and 150 by Tom Dunne

THE LIBRARY OF CONGRESS HAS CATALOGED
THE HARDCOVER EDITION AS FOLLOWS:
Provine, Robert R.
Laughter : a scientific investigation / Robert R. Provine.
p. cm.
Includes bibliographical references.
ISBN 0-670-89375-7 (hc.)
ISBN 0 14 10.0225 5 (pbk.)
1. Laughter—Psychological aspects. I. Title.
BF575.L3 P76 2000
152.4'3—dc21 00-038227

Printed in the United States of America
Set in Transitional 511
Designed by Jessica Shatan

To Helen

ACKNOWLEDGMENTS

A good-humored cast of characters contributed to this 10-year project of studying and then writing about laughter. I am particularly grateful to Yvonne Yong, Kenneth Fischer, Lisa Greisman, Tina Runyan, and Bernie Fischer, my energetic undergraduate research students, who contributed many of the observations and analyses presented in this book. I thank Kim Bard at the Yerkes Primate Research Center for making her chimpanzee charges available for studying and assisting in data collection.

I have intellectual debts to many teachers, students, and colleagues, but most of all to Viktor Hamburger and Rita Levi-Montalcini, who, many years ago, taught an embryonic graduate and postdoctoral student at Washington University (St. Louis) about the ways of the developing nervous system, and a life in science. They may not immediately recognize their influence, but it's there. And Barry Commoner freed me to graze at the interdisciplanary buffet of the university by providing me with a Junior Fellowship from his Center for the Biology of Natural Systems at Washington University. In more recent years, my colleagues at the University of Maryland Baltimore County have provided intellectual stimulation, friendship, and support.

I am deeply grateful to Robert Deluty and James Kalat, who read the entire manuscript and made many valuable suggestions. Steven

Pinker and Judith Rich Harris read major portions of the text and contributed to its readability. Thanks also to Charles Catania, Sheri Waldstein, Carlo Di Clementi, Marilyn Demorest, William F. Fry, Charles S. Harris, and Tom Benjamin, who commented on selected chapters. My editor, Rick Kot, offered good taste and sound advice throughout. I am also grateful to my agents, Katinka Matson and John Brockman, who helped to turn an idea into a book.

My greatest thanks go to my wife, pianist and musicologist Helen Weems, whose finely trained ear extends to prose. Helen read the entire manuscript, made invaluable comments at every stage of writing, and was a constant source of support, kindness, and laughter. This book is dedicated to her, with love and gratitude.

CONTENTS

Laughter

ONE

Laughter

An Introduction

he strangeness of laughter—as a behavior, and as a vocalization—is masked by its familiarity. Think about it the next time you walk through woods listening to the odd cries and calls of the creatures that live there: When you laugh, those creatures are hearing sounds that are just as odd and just as characteristic of our own species. Our raucous "ha-ha-ha"s give laughter its name. The verb "laugh," from the Old English *hliehhan*, is of onomatopoeic (sound-imitating) origin. Stripped of its variation and nuance, laughter is a regular series of short, vowellike syllables that are usually transcribed in English as "ha-ha," "ho-ho" or "he-he." These "words" are part of the universal human vocabulary, produced and recognized by people of all cultures. Laughter is instinctive behavior programmed by our genes, not by the vocal community in which we grow up.

In laughter we emit sounds and express emotions that come from deep within our biologic being—grunts and cackles from our animal unconscious. But what do these vocalizations signify? More than 2,000 years of the contemplation of laughter by some of history's great philosophers, writers, scientists, and physicians certify the importance of the question but provide only the vaguest of answers.

Laughter is a harlequin that shows two faces—one smiling and friendly, the other dark and ominous. Mardi Gras floats and sinister mechanical jokesters of old carnival fun houses mirror this duality—a volatile mix of gay and macabre that speaks directly to the emotional centers of our brain. Laughter can serve as a bond to bring people together or as a weapon to humiliate and ostracize its victims. Despots have rightly feared its power and have savagely repressed it. Plato thought that undisciplined laughter could threaten the state.

Given the social and emotional potency of the sound, our ignorance of laughter is remarkable. Amazingly, we somehow navigate in society, laughing at just the right times, while not consciously knowing what we are doing. Laughter is a "speaking in tongues" in which we're moved not by religious fervor but by an unconscious response to social and linguistic cues. Our brain has a masterful appreciation of the lawful relations between cues and response, but these rules are hidden from our conscious awareness. This book is the account of my 10-year search to discover those hidden laws.

My path of discovery led to valuable insights, many of them counterintuitive. Laughter, I found, is an ancient vocal relic that coexists with modern speech—a psychological and biological act that predates both humor and speech and is shared with our primate cousins, the great apes. Laughter provides some answers to the question of why apes can't talk and reveals critical steps in the evolution of speech and language. From tickling, an ancient stimulus for laughter, we learn of laughter's descent from the ritualized panting of rough-and-tumble play. The tickle is of interest in its own right, emerging from psychological obscurity to yield information about tactile communication and the perception of self.

Because laughter is largely unplanned and uncensored, it is a powerful probe into social relationships. It turns out, for example, that speakers laugh more than their audiences, that women laugh at men more than men laugh at women, and that laughter has more to do with relationships than with jokes. We will see that there is a pattern in social banter that once seemed to be only random "ha-ha-ha"s, and appreciate how laughter can make people seem warm, authoritative, cooperative, ineffectual, or just plain annoying. From studies of laughter's psychological consequences, we learn of laughter's therapeutic role.

The laugh tracks of comedy broadcasts provide insights into our tendency to respond to heard laughter with laughter of our own—an audience control technique that was exploited by the Emperor Nero and by the producers of *I Love Lucy.* If you doubt the contagiousness of laughter, consider that a 1962 epidemic in Tanganyika immobilized an entire school district for months. The knowledge of when and why people laugh provides novel applications, including how to maximize laughter in our lives and our environments, and how to evaluate its appropriateness and normalness. These discoveries will not enable you to laugh your way into a position of power or into a lover's arms, but they may help you to avoid laughing your way out of them.

My decision to pursue this topic was a professional step that raised some eyebrows, rather as if I had announced that I had decided to follow an unsavory guru, take up painting in middle age like Paul Gauguin, or become a used-car salesman. In the world of serious science, laughter is seen as a lightweight topic—an area lacking in clout and prestige. Its history has done little to enhance its reputation: Older studies of laughter tended to be known more for the enthusiasm of the researchers than for their rigor.

Like love, laughter has always hovered at the threshold of scientific scrutiny, more the province of the poet than the scientist. The challenge was to nudge it over that threshold and examine it as a problem in natural science. After 2,000 years of pontificating by philosophers, it was high time that we actually observed laughing people and described what they were doing, when they did it, and what it meant. This is the kind of work that should have been done 300 years ago. There may be no other area of human behavior where so many important questions remain unsolved and where it is still possible to do research with only pen, paper, and patience. This is research of the most democratic kind—based on ongoing human activity, neither arcane nor expensive, and open to rapid confirmation, extension, or challenge.

My research was conducted in the spirit of a zoological or anthropological field study, and an awareness that a fresh eye and ear were required to observe the familiar in new ways. Laughter had to be tracked down wherever it occurred—in bars, zoos, comedy clubs, acting classes, neurology clinics, malls, city sidewalks, class reunions, television laugh tracks, operas, lonely hearts ads, Pentecostal church services, and tickle wars. To study laughter meant grappling with some of psychology's knottiest problems—nature and nurture, consciousness, the perception of self, and the evolution of speech, language, and social behavior.

My quest for laughter was a reasoned extension of a 30-year search for what I consider the fundamentals of behavior, the bedrock of human nature. Unlike most students of laughter, I didn't come to the subject via the routes of philosophy, literature, or social psychology: My training is in behavioral neuroscience. I began my career in psychology with a naive faith that studying behavior at its most basic biological level—electrophysiology, neurochemistry, neuroanatomy, neuroembryology—would reveal truths inaccessible to more traditional approaches. In my early work, I investigated neurological mechanisms that pattern muscle contractions, the basis of all behavior, including laughter. I studied development in birds, focusing on the origins of flight and the effects of domestication, examining the development and evolution of specific movements, and searching for "behavioral fossils" of flight in living flightless birds. This work traced evolutionary changes in the neuromuscular system and revealed a mechanism that drives the natural selection of behavior.

My adventure in avian evolution led me to a similar search for archaic behavior in humans. What ancient behaviors are performed or lie dormant in modern humans? Do archaic acts provide the building blocks for the evolution of novel behavior, or does novel behavior, such as speech, spring forth de novo in a creationist flourish? What ancient behavior provides the species-wide foundation known as "human nature?" How has this behavior been transformed during the short evolutionary history of Homo sapiens? Armed with research on animal models and seeking bigger game (as well as relief from two decades of tedious procedures conducted mostly alone in windowless laboratories), I turned my attention to humans. My first foray into human behavior involved the mysterious act of yawning. Movement is movement, and no great inferential leap is required to shift from the wing-flapping of birds to the face- and body-flapping of yawning and laughing. Laughing is, in essence, a movement that produces a sound.

Why Laughter?

Some laugh aficionados have studied humor to show how it cures the common cold, increases creativity, or lifts depression. Although such insights are welcome, the present research agenda was established more on tactical grounds. As a research problem, laughter had just the right mix of significance and solvability. In science, there are no rewards for going into battle against the forces of ignorance if you are routed.

Laughter provides a lot of "scientific leverage," the ability to rigorously address a variety of questions. It is species-typical—that is, characteristic of our species—and stereotyped, meaning predictable in structure. Behavior is easier to comprehend if everyone performs it the same way, and neural mechanisms are easier to track down if everyone has them. Certainly the "ha-ha-ha" of laughter is more basic and rudimentary than the complexity and variability of speech, and it doesn't present the burden of having to understand its grammar. Another attraction of laughter is its remarkable contagiousness. Think of the last time you sat in an audience, laughing and letting waves of laughter wash over you. A pleasant experience—one of life's best. But consider now the primal nature of the animal chorus and the way the members of the audience synchronize their noises. How odd that we can't help laughing when we hear others laugh. This vocal chain reaction is evidence of a neurological process that detects and replicates laughter. Neural mechanisms have been proposed for the detection of the phonemes of speech, but laughter offers an even better chance of finding such vocalization-specific detectors. We are more likely to have evolved a mechanism to detect the simple, stereotyped, species-typical sounds of laughter than the more complex and variable sounds of speech. Unlike most social behavior, contagious laughter also offers an opportunity to move back and forth from the neurological to the social levels of analysis without trivializing either. As an added bonus, this entree into social neuroscience uses an easy-to-obtain and easy-to-maintain (no messy cages!) animal: Homo sapiens.

The Ethological Approach

How would a visiting extraterrestrial describe a group of laughing human beings to its relatives back home? What would the alien visitor make of these large, featherless bipeds emitting paroxysms of sound from a toothy vent in their faces? A reasonable approach would be to start by describing the most obvious aspects of the behavior: the characteristics of the animals emitting the sounds (such as their age and gender), the social context in which they emit the sounds and the rules that govern their expression, the structure of the sound itself and the mechanisms of its production and perception, and whether similar sounds are made by related species. Although the extraterrestrial's lack of familiarity with Earthling society would make our behavior difficult to interpret, our visitor would have one important advantage: a detached and unbiased perspective.

To Earthlings, this kind of naturalistic approach is known as ethology—a biologically oriented scientific discipline devoted to understanding what animals do, as well as how and why they do it. Ethology emerged from the European tradition of natural history, pioneered by Konrad Lorenz, Nikolaas Tinbergen, Karl von Frisch, and their nineteenth-century forerunner, Charles Darwin. Ethologists observe animals in their natural environments, treating behavior as an evolutionary adaptation, just like the anatomical structures of eye or wing. In searching for behavioral adaptations, ethologists focus on innate, instinctive behavior in a wide range of species. The ethological approach contrasts with that of many American behavioral scientists who more often focus on how the environment shapes behavior, using a few traditional animal models (e.g., rats, pigeons, humans). The present approach to laughter was inspired by ethological studies of animal calls and bird songs. Laughter, in this framework, can be regarded as an aesthetically and sonically impoverished "human song."

Focus on species-typical laughter is ideally suited to the ethological, natural history perspective, while it fits less easily into established niches of the more theoretically oriented psychosocial sciences. Laughter ranks high in *natural validity*—the significance of the act as measured by its frequency, amplitude, prominence, and social, psychological, and physiological consequences for our species. (Natural validity contrasts with *theoretical validity*, the behavioral relevance to issues posed by theories—for example, generative grammar in linguistics, cognitive dissonance or attribution theory in social psychology.)

My thinking about validity was sharpened in discussions regarding possible research funding with two research administrators in the area of speech science. One, a linguist, patiently explained that my proposed project "had no obvious implications for any of the major theoretical issues in linguistics." The other, a speech scientist, took another dismissive tack. Laughter, he told me, isn't speech, and therefore had no relevance to his agency's mission. I was being defined out of business, even though I was certain that studies of laughter would eventually illuminate theoretical issues dear to the hearts of both men.

The responses of these scientific bureaucrats were hardly unusual. Language-oriented books such as *Speaking* by Willem Levelt and *Listening* by Stephen Handel make no reference to laughter, a pattern of neglect typical of this literature. Laughter, it was clear, was an orphan behavior with a promising future, but currently without a theoretical home or means of support.

I began my investigation like a good ethologist (or extraterrestrial) by making observations of naturally occurring laughter, describing what laughter is, and where and when we laugh. Modest aims, perhaps, but achievable and a necessary foundation for what would follow. Without a knowledge of normal laughter, we can't know what hypotheses are worth testing, what is unique about laughter, and what might vanish in the socially impoverished environment of the laboratory. Moreover, unlike a lot of theory-driven work that might not outlast the fad that inspired it, systematic descriptions can serve as building blocks in many different theoretical systems, including those not yet devised. Good descriptions have a long shelf life. They are also conservative by nature and unlikely to contribute to the prevailing breeze that too often blows from the social sciences.

Sidewalk Neuroscience: The Neuropsychology of Everyday Life

A distinguished investigator once offered me advice about how to conduct a successful scientific career: "Dig an academic slit trench so deep and so narrow that there's only room for you." Many problems at the frontiers of science do indeed involve difficult, narrow, and deep issues accessible to only a few researchers with the resources and expertise to tackle them. Laughter, in contrast, is a poor subject for scientific trench warfare—it's too easy for newcomers with minimal training and resources to breach the barricades and storm the trenches. This makes laughter a rare instance of a significant scientific problem with frontiers near at hand and accessible to both professionals and amateurs. Laughter remains an unsolved problem because it has been hidden in plain sight. It has been overlooked because of the human tendency to neglect and undervalue the commonplace.

The accessibility of laughter as a problem makes it ideal for what I call *sidewalk neuroscience,* a low-tech approach to the brain and behavior based on everyday experience. Whether you follow in my scientific footsteps or simply read along, don't be put off by the primitive tools, simple methods, and behavioral focus. It's easy to be seduced by the irrelevant trappings of science and neglect the truly extraordinary in our midst. Too often we defer to experts, who tell us what is worthy and interesting and forget the natural curiosity about life and living that got the whole endeavor of science underway in the first place.

Treat this book as a field guide to the *terra incognita* of laughter, a source of tips about where to find laughter, how to study it, and what it means. You will not find a tidy series of experiments that drive inexorably (and with an intellectual flourish) to a Grand Unified Theory of Laughter. The laugh project is, instead, a catch-as-catch-can interdisciplinary work in progress. Some pieces of the laughter puzzle fit nicely into place, while others are parts of a yet unknown whole. Useful advice about the process is offered by embryologist Hans Spemann. "*I should like to work like the archeologist who pieces together the fragments of a lovely thing which are alone left to him. As he proceeds, fragment by fragment, he is guided by the conviction that these fragments are part of a whole which, however, he does not yet know. He must be enough of an artist to recreate, as it were, the work of the master, but he dare not build according to his own ideas. Above all, he must keep holy the broken edges of the fragments; in that way only may he hope to fit new fragments into their proper place and thus ultimately achieve a true restoration of the master's creation.*"

TWO

The Road Not Taken

Philosophical and Theoretical Approaches to Laughter

hilosophy is to science what alcohol is to sex: It may stir the imagination, fire the passions, and get the process underway, but the actual implementation may be flawed, and the end result may come up short. One by one, as scientific disciplines matured, they arose from their philosophical armchairs to set out on their own as the empirical sciences of physics, chemistry, biology, and most recently, psychology. But much of the literature about laughter is still mired in its prescientific phase where logic and anecdote, not empirical data, reign.

My years as a research scientist taught me respect—and a healthy skepticism—toward philosophical approaches to natural phenomena. All good science is built on a foundation of rigorous logic and solid scholarship, but science is much more than merely logic and scholarship. Without a critical eye on the natural world, science can lose its way, as it did during the European middle ages. An infusion of new data is necessary to guide our way and to prevent libraries of knowledge from deteriorating into musty memorials to long-dead authors. Even today, without the corrective influences of observation and hypothesis testing, we can be seduced into untenable positions by the masterful wordcraft, compelling logic, and prestige of our predecessors.

The task of reviewing the large prescientific literature on laughter, comedy, and humor is daunting. Much of the philosophically oriented literature on laughter from Plato to the present is what polite academics term "challenging." Be warned that, should you decide to pursue them, learned treatises on laughter can often induce a deadening of the brain, a glazing over of the eyes, and under the spell of especially earnest scholars, a slumping in one's chair.

The library's first lesson, however, is that the student of laughter is in good company. The undeniable importance of the topic is immediately revealed by the caliber of those who have sought to understand it, a group that includes Plato, Aristotle, Descartes, Hobbes, Kant, Schopenhauer, Darwin, Freud, and Bergson. The library's second lesson is that laughter must be a very difficult problem not to have yielded to these formidable intellects. The third, and most significant, lesson is that this literature is long on casual theorizing and short on empirical data, a fatal flaw that has impeded progress for over 2,000 years. My objective here is not to rehash this well-worn literature, which is the task for a different kind of book. I will provide instead only a brief sample of what has gone before to offer a taste of the nature and limits of an approach I have not taken, concluding with the transition from philosophical to empirical approaches to laughter.

The study of laughter dates from the first efforts of our species at self-contemplation and is documented in the most ancient philosophical writings. The earliest surviving theory of laughter is from

Plato (427–348 B.C.), one of the first and foremost of history's men of letters. Plato's considerable attention to laughter derived more from his fear of its power to disrupt the state than from delight with its practice. In his *Republic,* Plato discussed the negative consequences of abandoning ourselves to violent laughter. So as not to corrupt the young guardians-in-training of his ideal state, he went so far as recommending that literature be edited to delete mention of gods or heroes being overcome with laughter.

In *Philebus,* Plato suggests that what makes a person laughable is vice, especially the lack of self-knowledge among the relatively powerless. (The powerful were held to other standards.) Laughable people may see themselves as wealthier, more handsome, or smarter than they really are—a still viable description of many people whom we consider to be "jokes." (The workplace scenario of the popular *Dilbert* cartoon series by Scott Adams is peopled by a number of such self-important losers.) Plato also emphasizes laughter's association with pleasure and pain, making an analogy between relieving an itch by scratching and the laughable. Just as the presumed pain of the itch is relieved by the pleasure of scratching, the pleasure of laughter relieves the pain associated with gloating over friends' misfortune. (Plato assumed rejoicing at others' misfortune, an important cause of laughter, implied malice, which he thought to be painful.)

Plato's conclusion that laughter has a malicious element associated with the derision of our inferiors was developed further by many other philosophers over the centuries. Aristotle (384–322 B.C.) even contended that the gentle art of wit was a form of educated insolence, a position that would surely find support from Oscar Wilde and other masters of the form. Aristotle's treatment of laughter is based on indirect references (in his *Poetics*, *Rhetoric,* and *Nicomachean Ethics*), because the primary source is lost. (The missing book, in fact, is the one that caused the mischief in Umberto Eco's medieval mystery novel *The Name of the Rose.*) To Aristotle, the laughable is a subdivision of the ugly that does not cause injury or pain. The comic mask, for example, is distorted and ugly but is not pain-inducing (*Poetics*). Aristotle appreciated the effect of the unexpected in triggering laughter, an idea that was not pursued again for almost two millennia until the work of Kant and Schopenhauer.

Aristotle differs from Plato in believing that a little tasteful laughter is a desirable thing. Although one does not want to become a laughless boor, Aristotle cautions that we should nonetheless avoid too much of a good thing. "Those who go into excess in making fun appear to be buffoons and vulgar" (*Nicomachean Ethics*). Aristotle was also concerned with the use and abuse of laughter to persuade, discredit, and control. In *Rhetoric*, Aristotle cites the fifth-century Sicilian philosopher Gorgias (483–375 B.C.) who notes that jest is effective in killing an opponent's earnestness. Conversely, earnestness kills jesting.

Thomas Hobbes (1588–1679) builds upon Plato's and Aristotle's notion that laughter is associated with superiority over others. Given Hobbes's position in *Leviathan* that humanity is engaged in a constant struggle for power, it is not surprising that he awards laughter to the victor. In one of the best known quotations in the humor literature, Hobbes states that laughter is the expression of a "*sudden glory* (italics mine) arising from some sudden conception of some eminency in ourselves, by comparison with the infirmity of others, or with our own formerly" (*Human Nature*). Hobbes offers laughter as victorious crowing, the vocal equivalent of a triumphant flamenco dance stomped out on the chests of fallen adversaries. There is strong precedent for laughter as Hobbesian sudden glory, an expression borrowed by Barry Sanders for the title of his fresh historical treatment of laughter in literature and philosophy.

As we grapple with Hobbes and his predecessors concerning their view of laughter as derisive, it is helpful to remember that we are looking back to times when "good taste," "good manners" and "good form" were based on standards that were very different from today's—to times when the rich and powerful employed fools, physical deformity was a legitimate source of amusement, and the social elite might entertain themselves by visiting insane asylums to taunt the inmates. Even torture and executions were public events often conducted in a carnival atmosphere complete with snacks and refreshments. Upper and lower classes alike could revel in the knowledge that there were always those pathetic few who were less fortunate than themselves.

As social standards slowly evolved, the upper and then the lower classes adopted more contemporary views about what constituted proper targets of laughter, but even today we need not look far for evidence of laughter's darker, rowdier heritage. Contemporary comedians "kill" their audience with great jokes and "own" the crowd when their humor is well-received. In a vicarious form, we savor the crass, the grotesque and the vulgar in forms as diverse as Rabelais's *Gargantua and Pantagruel*, commedia dell'arte, Punch and Judy, The Three Stooges, and even in opera (see Chapter 4). What has become rare in the practice of daily life is perpetuated in literature and the arts.

We turn now to the more cognitive musings of Immanuel Kant (1724–1804), who, in the *Critique of Judgement*, states: "Laughter is an affection arising from the sudden transformation of a strained expectation into nothing" (i.e., with the delivery of the punch line of a joke, our expectation of how it will proceed vanishes). In this transformation, the strained expectation "does not transform itself into the positive opposite of an expected object—for then there would still be something, which might even be a cause of grief—but it must be transformed into nothing."

Arthur Schopenhauer (1788–1860), another earnest German with a cognitive position, further develops this idea, which has become known as the Incongruity Theory. To Schopenhauer, laughter arises from the perceived mismatch between the physical perception and abstract representation of some thing, person, or action, a concept that dates back to Aristotle. Our success at incongruity detection is celebrated with laughter.

Overflow or relief theorists offer a more physiological "hydraulic" perspective, viewing laughter as relieving an accumulation of nervous energy. Sigmund Freud (1856–1939) offers the best-developed relief theory in his *Jokes and Their Relation to the Unconscious*. He posits that all laugh-producing situations are pleasurable because they save psychic energy. Humor brings pleasure because it spares expenditure of feeling, comedy because it spares expenditure of ideas, and joking because it spares expenditure of inhibition. This excess, "spared," energy is relieved in the act of laughter, which serves as a kind of safety valve. As expected from the creator of psychoanalysis, Freud concludes that jokes are more than they seem. (Freud is credited with observing, "Sometimes a cigar is only a cigar." But can a joke ever be "only a joke?") Jokes, like dreams, have hidden benefits; both permit us to tap buried sources of pleasure because they permit access to the unconscious. As we will see in the next chapter, laughter is indeed a key to an "unconscious," but one very different than that envisioned by Freud.

Our tour down the philosophical path ends with the socially oriented contemplations of Henri Bergson (1859–1941) in *Laughter: An Essay on the Meaning of the Comic.* Although showing clear symptoms of philosopher's disease—an overly optimistic estimate of the power of naked reason and a dependence on anecdotal evidence—Bergson contributes the significant insight that all laughter is inherently social. To him, laughter loses its meaning and disappears outside of the context of the group, a position supported by current research. Bergson also suggests that laughter is a means of forcing compliance to group norms through humiliation ("ragging"). Most of us have, in fact, laughed at social outliers, although as we shall see, we probably do not consciously choose to act this way.

Curiously, Bergson suggested that an "absence of feeling" had to accompany laughter—to have an effect, the comic must fall on the surface of a "calm and unruffled" soul. This counterintuitive emphasis of the intellectual over the emotional neglects laughter's biological heritage in physical play, for we know now that the act of laughter is certain to have evolved before the cognitive and linguistic capacity to understand a joke, or even visual "slapstick" humor. Bergson adds that for something to be funny, it must be human; animals or inanimate objects become funny only in proportion to the degree that they remind us of something human. Bergson's examples include a puppet on strings and a jack-in-the-box. Our own experience confirms this, as we typically find the antics of humanlike capuchin monkeys funnier than those of such relatively inexpressive creatures as lizards, with their firmly sculpted facial visage. And cartoon figures from Bugs Bunny to Wile E. Coyote are funny largely because of their human appearance, motives, and actions. Less obvious is Bergson's suggestion that humans become funny in proportion to the degree that they can emulate machines. As he says, the laughable is "something mechanical encrusted upon the living." Is the visual comedy of Buster Keaton so effective because his expressionless face is incongruous with his otherwise human body and chaotic scenarios?

To portray the style and weakness of "science" by wit and anecdote, note Arthur Koestler's (*The Act of Creation*) attack on Bergson's position: "If we laugh each time a person gives us the impression of being a thing, there would be nothing more funny than a corpse." (Imagine, however, a mechanized corpse greeting guests at a "viewing" by sitting up in its casket and offering a prerecorded "Don't I look lifelike?" Is this not a thing acting like a person? Although this scenario is admittedly beyond ordinary experience, horrifying, and of questionable taste, it would fit nicely in a dark comedy about the funeral industry in the spirit of Evelyn Waugh's *The Loved One.*) John Morreall, a philosopher, challenges Bergson's idea that all laughter has a social basis by noting that "if I find a bowling ball in my refrigerator, I may find this incongruous situation funny, even though I do not see the ball as a person." Yes, but might you be laughing at whomever put the ball in the refrigerator?

The writings about laughter from which this sample was selected are bewildering in their mass and variety. There is something for every taste, including laughter in rhetoric and oratory (Cicero, Quintilian), literary analyses of what humor is and should be (Ben Jonson), how humor should be used, evolutionary theory (Darwin), archaic physiology (Descartes, Joubert, Spencer), and probes into intrapsychic events (Freud). Surveys of this literature are complicated because different authors of different professions communicate with different terminology in different languages over great gulfs of time. The most readily apparent feature of this theorizing is that most of it is really about humor or comedy (i.e., material that stimulates laughter), not laughter itself. This laughterless study of laughter continues to the present day.

Another problem with these very reasonable philosophical analyses becomes apparent when you ask, "What do I do next?," or "How do I decide between equally plausible positions?" Too often, there is no "next," no critical experiment or observation that might bring insight and understanding, or resolve disputes between opposing camps. There is only an endless battle between dueling essayists and the churning of the same familiar material. An even more serious flaw of philosophical analysis is its tacit and unfounded assumption of the "rational person" hypothesis—namely, its common sense but incorrect premise that the decision to laugh is a reasoned, conscious choice. Common sense does not always serve us well in the surprising and sometimes counterintuitive world of laughter. Philosophical inquiries also fail because they are too far removed from the phenomenal world they seek to explain.

In contrast to the long tradition of philosophical analysis, the history of empirically based laughter and humor science is little more than 100 years old. Laughter and humor research appeared gradually and sporadically, roughly paralleling the emergence of experimental psychology in the late nineteenth century. With the turning inward of the scientific method to study mental life came incidental empirical reports that joined the always substantial popular literature about laughter and humor. G. Stanley Hall, a founder of American psychology, contributed a questionnaire-based study of tickle in 1897. Other scattered contributions to laugh science included an introspective analysis of the humorous (Martin, 1905), memory for funny material (Heim, 1936), laugh-provoking stimuli (Kamboropoulou, 1930), children's laughter (Kenderdine, 1931; Ding and Jersild, 1932; Justin, 1932), and studies of laughter development

(Washburn, 1924; Wilson, 1931), a line of research that began with the anecdotal reports of Charles Darwin (1872, 1877). Separate but significant lines of inquiry were pursued in the clinical case studies of neurology (Wilson, 1924) and psychiatry (Kraepelin, in Defendorf, 1904; Bleuler, 1911, 1916). As the study of laughter and humor matured, systematic observation and experimentation gradually displaced introspection and the collection of anecdotes. But the sparse research never reached critical mass. During most of the twentieth century, isolated researchers were left to their own devices, toiling alone.

The pace of research accelerated during the 1970s and 1980s when organizational and scholarly advances brought a semblance of structure and direction to professional and amateur students of laughter, a motley horde of philosophers, physicians, nurses, anthropologists, linguists, psychologists, physiologists, English professors, cartoonists, comics, and clowns. These decades featured the formation of the International Society of Humor Studies and its journal *Humor*, the now yearly International Humor Conferences, and the publication of books that assembled previously scattered work. Such books quickly became "classics" and have aged rather well, a measure of the leisurely pace of the field. Enthusiasm for laughter/humor studies was further stimulated by the publication in 1979 of Norman Cousins's book *Anatomy of an Illness as Perceived by the Patient.* Cousins's evangelism for the health benefits of laughter triggered a movement that continues to this day.

Like the work of the philosophers, however, laughter research was not born whole—once again, laughter itself was often missing. Most laughter research neglects laughter and its occurrence in everyday life, focusing instead on a variety of corollary issues about humor, personality, social dynamics, and cognition. A few representative studies illustrate the point.

Hans Eysenk asked why one person likes crude jokes, while another prefers more subtle comic fare. On the basis of ratings provided by his subjects, he concluded that extroverts preferred sexual and simple jokes, while introverts preferred nonsexual and more complex humor. In another study, Herbert Lefcourt examined the relation between humor style and "locus of control," a person's perception of whether he was in control of his own destiny ("internalizers") or the victim of circumstance ("externalizers"). Internalizers used all types of humor more than the externalizers, who engaged mostly in "superiority" or "tension relief" forms. Other psychological studies have evaluated the role of incongruity in jokes or shown that we laugh more when bad things happen to obnoxious than to pleasant people, evidence of the aggressive roots of humor.

As interesting and worthy as such studies may be, they again only peripherally concern laughter, focusing instead on humor and ancillary issues about theories of personality (e.g., introversion/extroversion, locus of control), social dynamics (e.g., social facilitation, deindividuation, release constraint mediated by imitation), laughter's precursors and consequences, health benefits, what makes jokes funny (e.g., incongruity), or why we laugh at jokes (e.g., aggression). This book redresses this historical imbalance by putting the act of laughter front and center.

Good science is simplification, and in this spirit we begin the study of laughter by focusing on laughter itself, instead of its correlates, applications, or theoretical implications. And like good journalists, we start with the "who," "what," "when," and "where" of laughter, the small but sure empirical steps that guide us, haltingly, toward the core of our social being. We avoid the traditional introspective and commonsense approaches that are ill-equipped to reveal the unconscious, instinctive mechanisms of laughter. Instead, we search for laughter wherever it is found, and follow the empirical trail wherever it leads, from the stage of grand opera to tickle frolics. Along the way we consider the evolution of the nervous system and the organs of vocalization—an essential but missing step of previous approaches to laughter. In the primal panting laughter of chimpanzees, for example, we discover why humans can talk and other apes can't, and why walking upright on two legs was necessary for this transformation. We also learn why tickling and rough-and-tumble play teach us more about the roots of laughter than their evolutionary by-products of joking and stand-up comedy, the behaviors that captured the attention of so many philosophers and social scientists. The strongest argument for the present, naturalistic approach to laughter is that the most useful approaches and the most exciting discoveries were unanticipated at the outset—they were suggested by observations of laughing people.

THREE

Natural History of Laughter

rom philosophers we have learned a lesson about laughter that they did not intend to teach—namely, that intellectual prowess, however formidable, is limited in what it can tell us about this tantalizing behavior. To learn more about laughter we must stop talking about it and start listening to the ultimate experts, laughing people. This chapter traces the first steps in a 10-year search to understand what laughter is, when we do it, and what it means. These early investigations were an improvisational scramble because there were few precedents about how to study ongoing laughter in everyday life. Most previous research about "laughter" did not, in fact, concern the act of laughter, and was conducted in laboratories or used paper and pencil tests. My adoption of a naturalistic, descriptive tactic was the basis for all that follows, including the discovery of several new phenomena of laughter. As we will see, careful observations of ongoing behavior can lead to startling and often counterintuitive insights into the nature and neurological control of laughter. But the wisdom of this approach was not obvious at the outset. Learning how to learn about laughter held lessons of its own.[1]

"Making tiger soup is easy. First you get a tiger," goes a popular joke. My first efforts to find laughter to study were rather like making tiger soup. With naive enthusiasm, I set out to find laughter, the essential ingredient. But for something so apparently robust and common, laughter proved surprisingly fragile and illusive. Bring laughter under scrutiny, and it vanishes.

I began with the assumption that the best way to obtain laughter for study was to invite individuals or small groups of two or three people into my laboratory and entertain them with audio and video recordings of comedy performances. My subjects would surely surrender to the comic powers of a Rodney Dangerfield or a Joan Rivers. If not, then George Carlin's classic "Seven Dirty Words" routine, the *Broadcast Bloopers* album, or some of the very funny early episodes of *Saturday Night Live* would certainly turn the trick. But nothing seemed to work. My comic virtuosi elicited only a few grudging chuckles from the lab-bound subjects. I was surrounded by laughing people who would go stone sober when brought into the laboratory.

What seemed like a "can't-fail" project was failing. This was a humbling experience, for obtaining a sample of laughter does not qualify as what many people would consider rocket science. At this point, I learned two lessons. First, any research dealing with the perverse social tendencies of human beings *is* rocket science. Second, *the subject is always right.* These "lessons" should be elevated in status to "principles," ranking with other universal truths like, "There's no such thing as a free lunch," and "If anything can go wrong, it will go wrong."

[1]Hard-won insights about where to find laughter and how to increase laughter in your life are provided in the Appendix.

My failure to elicit laughter in viewers of comedy performances was actually a success in disguise. By not laughing, my subjects were announcing that laughter is a social behavior that virtually disappears in isolated people being scrutinized in a laboratory setting. Henceforth, I would get out of the laboratory and let laughing people in public places tell me what is important about laughter through their normal behavior. Although this approach seems obvious in retrospect, hardly anyone in the long history of laughter studies ever observed laughing people in the field.

In the spirit of Jane Goodall heading out to Gombe Stream Preserve to study chimpanzees, three undergraduate assistants and I set forth on an urban safari to study humans in their natural habitat of shopping malls, city sidewalks, and the university student union. We were surprised how much could be learned by simply listening to people speak and laugh. What began as a last-ditch effort to rescue a foundering project led to exciting and unanticipated discoveries.

Anatomy of the Laugh Episode

Our approach to studying laughter was straightforward. We eavesdropped on the conversations of anonymous laughing people in public places. No one seemed aware of our interest, although a large and aggressive woman at a local mall once accused me of being a store detective. The challenge was in making sense of what we were hearing. At first, the "blah-blah-blah-ha-ha-ha-blah-blah-blah" of conversational laughter sounded like a foreign language, lacking structure and meaning. Several weeks of trial and error were required to discover what now seems obvious.

Imagine yourself at a large, successful party with plenty of good food, good drink, and good companionship. It's a noisy affair, buzzing with simultaneous conversations and punctuated with laughter. John is relating a not very entertaining story, but is rewarded in his efforts by the hearty laughter of an attentive group of three or four. Jennifer chimes in with her own comments, but with less effect—her giggles are accompanied by reserved nods and smiles, but no laughs. Variations of this social give-and-take are repeated many times during the evening. But who actually is laughing and why? Is there order in this apparent chaos of chatter and laughter? The challenge here is to parse the stream of conversation and laughter into manageable, quantifiable units.

The first advance in understanding conversational laughter was in conceiving the *laugh episode.* A laugh episode consists of the comment immediately preceding laughter, and all laughter occurring within one second after the onset of the first laughter. Two laugh episodes were just described in the party scenario. In one, John spoke and his audience laughed. In the other, Jennifer both spoke and laughed, but her audience remained silent.

The second advance was in distinguishing between *speaker* and *audience laughter.* In conversation, the speaker is the person talking, and the audience is the person or persons being addressed. Audience laughter has been the almost exclusive concern of previous researchers, while speaker laughter has gone largely unrecognized. At first, I, too, neglected speaker laughter. Consider the above example in which Jennifer the speaker laughed, while her audience did not. When observing who laughed in response to speaker comments, I was bothered by the fact that the speaker often laughed and was, in fact, sometimes the only person who laughed—a complication that was creating chaos in my scoring system. I then realized, however, that the speaker was part of the social unit and had to be included in any analysis. My slowness to appreciate the speaker's laughter was attributable to the pervasive and inappropriate model of stand-up comedy in laughter research, in which a usually nonlaughing (deadpan) comedian tells jokes to a receptively laughing audience.

The third advance in understanding laughter was in recognizing *gender differences* in laugh patterns. Gender is a fact of any human interaction, and recognizing gender differences is essential to understanding patterns in conversational laughter.

The strategy for studying conversational laughter was now taking shape. When laughter occurred, we would record (1) who laughed: the speaker (S), audience (A), or both; (2) the gender of speaker and audience, and (3) the prelaugh comment. For example, if John the male speaker (S_m) told a story to Jennifer his female audience (A_f), and both laughed within one second, a circle would be drawn around both speaker and audience, (S_m A_f), denoting that they both laughed. If Jennifer commented that John had onion dip on his tie, and Jennifer but not John laughed, (S_f)A_m would be recorded, signifying that only the female speaker laughed. To simplify our analysis, we examined the smallest possible social group, the dyad, which consists of only two people.

The fruit of our yearlong eavesdropping mission was 1,200 laugh episodes that were sorted by speaker and audience, gender, and prelaugh comment. These hard data forced me to attend to new laugh phenomena, and prompted my reluctant metamorphosis from neuroscientist into social psychologist.

Gender in Laughing, Speaking, and Listening

There are no gender-free human encounters. When someone laughs, someone else is usually present, and the gender of that person must be taken into account to understand the social dynamics of laughter. Linguist Deborah Tannen pointed out the significance of gender differences in conversational styles in her best-selling book *You Just Don't Understand.* The gender differences in laughter may be even greater, the evidence for which we will now consider. (The percentages in the table on page 28 do not add up to 100 percent because either the speaker, the audience, or both may laugh during a given episode.)

While tabulating the data, I found that speakers laughed more than their audiences. Nothing in the audience-oriented literature about laughter or humor suggested such a result. When I totaled speaker (S) and audience (A) laughter across all four possible gender combinations, speakers were found to laugh 46 percent more than their audiences. The effect was even more striking when gender was considered. The speaker/audience difference was greatest when females were conversing with males (S_f A_m), a condition in which fe-

Dyad	Episodes	Percentage Laughing Speaker	Audience
$S_m A_m$	275	75.6%	60.0%
$S_f A_f$	502	86.0%	49.8%
$S_m A_f$	238	66.0%	71.0%
$S_f A_m$	185	88.1%	38.9%
overall	1200	79.8%	54.7%

males produced 126 percent more laughter than their male audiences.

When a female colleague first saw these results showing the high levels of female speaker laughter, her response was, "Oh, my God, the stereotype of the giggling female has been confirmed." Another female colleague put a different spin on these data, noting that "when dealing with males, there is so much more to laugh at," an argument weakened by the finding that female speakers laugh only slightly less at other females than at males. Female speakers are enthusiastic laughers whoever their audience may be. Male speakers are pickier, laughing more when conversing with their male friends than with an audience of females. The least amount of speaker laughter occurred when males were conversing with females—this $S_m A_f$ grouping was the only one of the four gender categories having less speaker than audience laughter. As audiences, both males and females were more selective in whom they laughed at or with than they were as speakers—neither males nor females laughed as much at female as male speakers. In summary, *females are the leading laughers, but males are the best laugh getters.*

The male superiority in laugh-getting develops early in life. In England, Hugh Foot and Anthony Chapman observed that among children viewing cartoons, girls laughed more often with boys than with girls, and girls reciprocated boys' laughter more often than boys reciprocated girls' laughter.

The gender pattern of everyday laughter also suggests why there are more male than female comedians. Sexism and bias in show business may not be the only explanation, for comics Rosie O'Donnell and Phyllis Diller may actually have to work harder for their laughs than Rodney Dangerfield or Jerry Seinfeld. As a male, Rodney Dangerfield gets more respect than he claims. The high proportion of male comics may also be related to the number of trainees in the pipeline. Males engage in more laugh-evoking activity than females, a pattern that may be universal. In a cross-cultural study of children's humor in Belgium, the United States, and Hong Kong, males were the principal instigators of humor, and this tendency was already present by six years of age, when joking first appears. Think back to your own childhood days and recall who was your class clown—most likely it was a male.

While males are the leading jokesters, and females are the leading laughers and consumers of humor, what is actually being communicated in their noisy displays? Might laughter be performed by a subservient individual, most often a female, as a vocal display of compliance, subordination, or solidarity with a more dominant group member? Insights about the social role of laughter can be gleaned from laugh patterns of people holding different social rank within a group.

In a rare and enlightening naturalistic study, sociologist Rose Coser found a strong relation between humor production, the target of humor, and the professional status among staff at a psychiatric hospital. (The analysis focused on laughter that was a response to an anecdote or other "intended provocation.") During staff meetings, the senior staff (psychiatrists) most often made junior staff (residents) the target of their witticisms. The junior staff did not reciprocate, most often targeting instead patients or themselves, a pattern also typical of the lower-ranking paramedical staff (psychologists, social workers, sociologists). Not once did the junior staff target a senior staff member present at a meeting. Consistent with this finding of *downward humor,* the paramedical staff never made any member of the psychiatric staff (senior or junior) the butt of their humor. Humor apparently had high social costs only senior staff could afford. The average witticisms per staff member were 7.5 for senior staff,

5.5 for junior staff, and only 0.7 for the lowly paramedicals. These values are especially striking because junior staff did most of the talking at meetings. Also notable was the male contribution of 96 percent (!) of witticisms, despite substantial female representation in all staff ranks. Although wit was costly, laughter was cheap—women, for example, seldom joked but were enthusiastic laughers.

Are these laugh patterns stable over time and social circumstance? For example, is a person's laugh pattern fixed, or is it conditional, so that it changes with promotions or demotions in social rank? It is likely that a person assumes a variety of state-specific laugh patterns, adopting the one most appropriate for a particular social situation (e.g., with professional colleagues, friends, children). A stern boss may be a barrel of laughs when cavorting with old school chums.

Laughter's role as a signal of dominance or subservience can be tested by observing whether male underlings switch to a more typically female laugh style when conversing with female bosses (e.g., laughing a lot as both speaker and audience). Another approach would be to contrast the laugh behavior of such powerful women as Hillary Clinton and Margaret Thatcher with that of their male subordinates. A change in power status probably brings a shift in laugh pattern.

Ethnological studies document the flexibility and strategic use of laughter. In southern India, men belonging to a lower caste giggle when addressing those of a higher caste. Other aspects of "self-humbling" are well developed among Tamil villagers of low caste (Harijan), but are exercised only when dealing with powerful persons of higher caste. When dealing with a landlord, for example, a Harijan may giggle, speak with unfinished sentences, mumble, appear generally dim-witted, and when walking, shuffle along. Yet this same Harijan may suddenly become shrewd and articulate when dealing with less powerful people.

Similar self-effacing behavior, including buffoonery, is practiced by the Bahutu in the presence of their caste superiors in Central Africa. And the women of many societies worldwide exhibit various forms of "self-humbling" in the presence of men (i.e., lowered eyes, shy or embarrassed silence). The "prosodic" (nonlinguistic sonic quality) of speech also bears social status information about the speaker and audience. Men and women of the Tzeltal (Mexico) and Tamil use high-pitched/falsetto voices to convey self-humbling. Presumably, the high-pitched voice shows deference and is less threatening because it is characteristic of women and children. Among the rural Tamil, male and female Harijan (low-caste) speakers address high-caste powerful persons with thin, high-pitched voices. A higher pitch is also used within the caste to show deference or power asymmetries, as when a daughter is making a request of her father.

Let us now return to our friends at the party and continue the exploration of laughter in organizational politics. Consider Jennifer, a woman of high intelligence and good judgment, who giggles a lot as speaker and audience. Will she unfairly be passed over for a management position in her company because of her ubiquitous laughter, a decision her detractors may falsely attribute to her lack of leadership skills? Ironically, neither Jennifer nor her detractors would recognize the role that laughter might play in her evaluation. Does her giggly laughter disqualify her from positions of authority? Or would Jennifer's promotion to a management position bring with it a change to more role-specific laughter? I predict that Jennifer's laugh pattern would shift to match her new, more responsible position even though she would be unaware of the transformation.

Evaluate your own experience with organizational laughter. Is sharing laughter with subordinates incompatible with positions of authority? Have you ever encountered a leader of high authority who has a giggle? How many really funny generals are there? Would such a person be considered a "serious" or "formidable" member of an organization? How many presidents open their State of the Union speeches with one-liners, in the manner of banquet speakers on the rubber chicken circuit? Someone who laughs a lot, and unconditionally, may be called a "ditz," or a "good ol' boy," but seldom "boss," or "president." The late American legislator Morris Udall recognized this issue when he titled his autobiography *Too Funny to Be*

President. John F. Kennedy was unusual among U.S. presidents in having both a presence of command and an excellent sense of humor.

In Search Of: Laughter and Humor Requests in the Personals

Given the differences in male and female laugh patterns, is laughter a factor in human meeting, matching, and mating? Can we learn to laugh our way to social and sexual success? These possibilities are particularly intriguing, because we have little awareness or conscious control of our laughter. My exploration of the sexual politics of laughter uses a technique even less exotic than field trips to the local mall—these data were ready-made and published in the personal ads of newspapers.

> *HEALTHY, HAPPY, wholesome, sexy, funny, smart, playful, spiritual, pretty, blond 5'7"/132 ISO same qualities in tall, dark hunk, 45+.*
>
> San Diego Union-Tribune

> *SINGLE white professional male, 6'1" 200 lbs., blue-eyed blond, enjoys all sports, outdoors, movies, laughter, ISO attractive, athletic SWF 21–26, with good sense of humor.*
>
> Chicago Tribune

What do men and women want in a lover or life partner? Is sense of humor and funniness part of this human equation? Many of us value friends who have a lighthearted view of life and avoid those who radiate a dark aura of pessimism and depression. But do we consciously seek out partners who love to laugh and make us laugh? Personal ads deserve careful consideration because a lot of thought probably goes into their composition. After all, the seekers had better be careful about what they ask for, because they might get it. And what is being sought in this catalogue of human desire? A few people have simple generic needs ("male," "female"), but most are more specific about age, height, body build ("fit," "hard body," "proportionate," "full-bodied"), marital status, personality, sexual orientation, religion, drinking, drugs, smoking, and race. (One discriminating person sought a partner who was "ebola free.")

Social scientists have identified consistent trends in this human marketplace. Men more often *seek* physical attractiveness and *offer* financial resources than women. Conversely, women are more likely than men to *offer* attractiveness and *seek* financial resources. Males and females show a clear and convenient complementarity in what is being offered and sought, a necessary condition for successful deal making. Economists will find a lot to like here. Another trend in the ads may not surprise females who are presumably more perceptive in such matters—more males than females focus on physical characteristics (e.g., height, weight, age, eye and hair color), whereas more females consider personality and psychological traits (e.g., intelligence, sensitivity, caring, spirituality, maturity). Among advertisers mentioning age, males more often seek younger partners and females older ones.

But what is the currency of laughter in the human marketplace of personal ads? To find out, I perused 3,745 ads placed by heterosexual males and females in eight major, mainstream, geographically dispersed American newspapers on Sunday, April 28, 1996. A single date was examined to avoid repeated ads. The papers were the *Baltimore Sun, Boston Globe, Chicago Tribune, Cleveland Plain Dealer, Houston Chronicle, Miami Herald, San Diego Union-Tribune,* and *Washington Post.* Overall, men took out 23 percent more ads than women (2,065 versus 1,680), but the proportion ranged from the *Houston Chronicle* with 21 percent more female ads to the *Cleveland Plain Dealer* with 28 percent more male ads. (Homosexual males and females were not considered, because they contributed a relatively small number of ads and deserve a separate study of their own.)

Laughter or laughter-related behavior (e.g., "humorous," "sense of humor," "funny," "witty") was mentioned in about one-eighth of the total ads (13 percent, or 478 of 3,745). Ads seeking or offering "fun" or "having fun" were not counted because such general descriptors may not involve laughter. Females were much more likely (62 percent more) than males to mention laughter in ads, a reflection of women's greater relative concern with personality and psychological traits. The plot thickens when laugh citations were divided into *seek* and *offer* categories for males and females. Advertisers were considered to seek laughter in a desired partner if they requested either laughter itself ("loves to laugh," etc.) or laugh-producing behavior ("sense of humor," "funny," etc.). The advertisers were considered to offer laughter if they either mentioned their own tendency to laugh, sense of humor, or appreciation of laughter, even someone else's.

The evidence is clear. *Women seek men who make them laugh, and men are anxious to comply with this request.* Women sought laughter (13 percent) more than twice as often as they offered it (5.7 percent). In contrast, men offered laughter (6.5 percent) about a third more than they sought it (4.9 percent). The most common form of bartered laughter was "sense of humor." Men offered it and women sought it, although not in identical proportions. The complementarity of male and female laughter requests is striking because laughter is under minimal conscious control and neither sex may be aware of the gender differences in laughter. Both sexes unknowingly comply with the demands of their instinctive script.

What is the nature of the laughter offered by men and sought by women? As we have seen, observations of actual behavior indicate that men laugh much less than the women with whom they are conversing. But laughless males are an unlikely concern of women who are really requesting men who make *them* laugh. And the men who offer laughter aim to stimulate it in their female partner, not to laugh themselves. These men may revel in the chuckles of a female companion, a measure of her pleasure and recognition of his acceptance and status. Such "funny" men are likely to pass the female laugh test whether they actually laugh or not. Given the previously considered relation between laughter and social status, the desire by women for men who make them laugh may be a veiled request for dominant males. Men who pass the audition for dominance are acknowledged with women's laughter. However, women may reward dominance more than men reward submission—men are less likely to seek laughter in their personal ads than women.

Laughter-seeking by women was reported by Karl Grammar and Irenäus Eibl-Eibesfeldt in a study of spontaneous conversations between mixed-sex pairs of young German adults who were meeting for the first time. The more a woman laughed aloud during these encounters, the greater was her self-reported interest in seeing the man again. The man's own laughter did not indicate interest in the woman, but he was interested in meeting a woman who laughed heartily in his presence. (Simultaneous male and female laughter did, however, predict mutual interest.)

Imagine a conversing couple standing close and gazing intently into each other's eyes, with the man's indistinct "mumble-mumble-mumble" repeatedly punctuated by the bright sound of female laughter, and a lot of shared laughter. My data suggest a positive prognosis for this relationship. The prospects are less favorable for a relationship featuring a lot of unreciprocated male laughter—the couple may be standing farther apart, and the female may be glancing around, signaling lack of interest in her companion, providing an invitation for outsiders to join them.

In the above scenarios, *the laughter of the female, not the male, is the critical positive index of a healthy relationship.* Guys can laugh or not, but it's best that their woman is getting her yuks in. Pop diva Cyndi Lauper advanced this hypothesis in her perceptive song "Girls Just Want to Have Fun." But on the basis of evidence stated earlier, Madonna deserves an equal hearing for her "Material Girl" hypothesis. Who is right? Probably both. Current research of the personal ads suggests that a woman will most enjoy the company of a man who loves laughter (at least that of his girlfriend), has a good job, and some ready cash in his pocket.

Laughter Punctuates Speech

Our lives are filled with a torrent of words that are punctuated by brilliant eruptions of laughter. Amazingly, we somehow make sense of this word salad and speak, laugh, and listen at just the right time. This triumph is all the more remarkable given the biological equipment at our disposal. Those viewing humans as the masterwork of creation should consider our jury-rigged contraption for speech, no part of which evolved expressly in the service of speech. A bioengineer conceiving such a device would be sent back to the drawing board, or out the door.

Consider the facts: We eat, drink, breathe, belch, vomit, and talk through the same orifice, a plumber's nightmare with sharp teeth at one end. Is this any way to design an organism? We must stop breathing to swallow—or suffer severe consequences. (Babies can simultaneously breathe and nurse during the early months of life.) Likewise, breathing must cease when we speak. And we must choose whether we would rather speak or eat. The only way we are able to live with such an inefficient mechanism is through an ingenious time-sharing program. We reflexively stop breathing when we swallow and rarely breathe in mid-sentence.

Like much in nature, we get by not because of superior organ craftsmanship (Mother Nature is a notoriously sloppy workperson), or elegance of design (remember the troublesome wisdom teeth and the vestigial appendix), but through some resourceful biologic innovations, this time involving the neurological wetware of our central nervous system. Our brain choreographs this complicated ballet of breathing/eating/drinking/talking so well that we are scarcely ever aware of it—a triumph of biologic error control.

Here we examine the temporal patterning of laughing and talking by observing the placement of laughter in the speech stream. Because the neurological processes that produce laughter and speech converge on a single mechanism of vocalization, a study of vocal behavior reveals how we deal with this confluence of potentially conflicting vocal acts. Does laughter or speech win this competition for vocal dominance? Is laughter randomly inserted into speech, or is laughter inserted only at phrase breaks or pauses in speech, which would suggest evidence of speech dominance? And if there is a lawful, nonrandom relation between laughter and speech, what is its organizing principle? *Laughspeak,* a form of blended, laughing speech that communicates emotional tone, is qualitatively different from the classical ha-ha-type laughter considered here and was excluded from this analysis.

The inquiry into the relationship between laughter and speech is based upon the same file of 1,200 laugh episodes that was considered in a previous study, except the focus here is upon the placement of laughter in the prelaugh comment. Of particular interest is whether laughter interrupts the phrase structure of speech. The results are gratifyingly clear.

During conversation, laughter by speakers almost always follows complete statements or questions. Laughter is *not* randomly scattered throughout the speech stream. Speaker laughter interrupted phrases in only 8 (0.1 percent) of 1,200 laugh episodes. Thus, a speaker may say, "You are going where? . . . ha-ha," but rarely "You are going . . . ha-ha . . . where?" This strong and orderly relationship between laughter and speech is akin to punctuation in written communication and is termed the *punctuation effect.*

Speech has a clear priority in gaining and maintaining use of the single vocalization channel because it's seldom interrupted by laughter. Although people may recall instances when they or others "break up" in laughter, and have difficulty finishing jokes or stories, such interruptions are actually quite rare. We should avoid making exceptions the rule, a common tendency that thwarts the discovery of basic mechanisms and principles.

The occurrence of speaker laughter at the end of phrases indicates that a lawful and probably neurologically based process governs the placement of laughter in speech. The temporal segregation of laughter and speech is evidence that different brain regions are involved in the expression of cognitively oriented speech and the more primitive, emotion-laden vocalization of laughter. (Additional neurological evidence for this division of labor is presented in Chapter 8.) During conversation, we switch back-and-forth between the speech and laughter modes. During speech, the dominant speech-producing region inhibits that producing laughter. If laughter is triggered during speech, its expression must await the passing of phrase-sized chunks of speech-related inhibition.

The punctuation effect holds for the audience as well as for the speaker, a surprising result because the audience could laugh at any time without speech-related competition for their vocalization channel. No audience interruptions of speaker phrases were observed in our 1,200 laugh episodes. It's unclear whether the punctuation of speech by audience laughter is cued directly by the speaker (e.g., a postphrase pause, gesture, or laughter), or by a brain mechanism similar to that proposed for the speaker that maintains the dominance of language (this time perceived, not spoken) over laughter. *The brains of speaker and audience are locked into a dual-processing mode.* (Further evidence of the synchronization of the brains and behavior of speakers and audiences comes from the remarkable phenomenon of contagious laughter considered in Chapter 7.)

Given laughter's punctuation of speech, it's logical to ask what kind of punctuation laughter represents—a period, a question mark, an exclamation, or something else. Although the imposition of specific punctuation on the rich, multimodal conversational flow is necessarily subjective and may attribute more formal structure to conversation than warranted, it is sufficient to show that laughing is not an exclusive consequence of a particular type of comment. Laughter by speaker or audience followed statements in 84 percent, and questions in 16 percent of laugh episodes. The precise proportion is less significant than the demonstration that laughter commonly follows both statements and questions.

Observers awarded an exclamation to 41 percent of sentences (statements and questions) preceding laughter. Laughter often followed such statements as "What a shirt!" or "You did what?!" Exclamatory sentences, a highly subjective categorization, ended on a crescendo, change in intonation, or other attention-getting nuance associated with elevated emotional tone.

The first report of a defect in the punctuation process may have been made in *Thinking in Pictures,* Temple Grandin's fascinating, autobiographical account of life with autism. She notes that "when several people are together and having a good time, their speech and laughter follow a rhythm. They will laugh together and then talk quietly until the next laughing cycle. I have always had a hard time fitting in with this rhythm, and I usually interrupt conversations without realizing my mistake. The problem is that I can't follow the rhythm." It's significant that Grandin is struggling with an ability that nonautistic individuals mindlessly perform with remarkable accuracy.

The punctuation effect finds application in joke telling. Master comedians are aware that success lies as much with the presentation of a joke as with the joke itself. A critical part of joke telling is timing, the pace of storytelling, the setting up, and delivery of the punch line—and most important to our present story, the pause that follows the punch line.

Some comedians, like Groucho Marx, work quickly, delivering a rapid-fire barrage of jokes, while others, like Jack Benny, proceed at a more leisurely pace, in the tradition of storytelling. But a critical feature in the style of all stand-up comedians is a pause after the delivery of the punch line, during which the audience laughs. The comic usually signals the onset of this critical pause with marked gestures, facial expressions, and altered voice intonation. Jack Benny was known for his minimalist gestures, but they were still discernible, and worked wonderfully. A joke will fail if the comic rushes to his next joke, providing no pause for audience laughter (*premature ejokulation*)—this is comedy's recognition of the power of the punctuation effect. When the comic continues too soon after delivery of his punch line, he not only discourages, and crowds-out, but neurologically *inhibits* audience laughter (*laftus interruptus*). In show-biz jargon, you don't want to "step on" your punch line.

Nothing to Joke About: What People Say Before They Laugh

If you want people to laugh, you tell them jokes. Right? Well, not according to our survey of what 1,200 people said immediately before they laughed. Jokes will work, but people laugh more often after such innocuous lines as "I'll see you guys later" or "Are you sure?"—not exactly knee-slappers. Our fieldwork failed to discover The Mother of All Jokes or even her next of kin. In fact, most laughter did not follow anything resembling a joke, storytelling, or other formal attempt at humor. Only about 10 percent to 20 percent of prelaugh comments were estimated by my assistants to be even remotely humorous. Peruse the sample list of 25 Typical Prelaugh Comments and make your own decision about their humorousness.

Table 1

25 Typical Prelaugh Comments

Typical Statements

I'll see you guys later.
Put those cigarettes away.
I hope we all do well.
It was nice meeting you too.
We can handle this.
I see your point.
I should do that, but I'm too lazy.
I try to lead a normal life.
I think I'm done.
I told you so!
I was completely horrified!
There you go!
I know!
Must be nice!
Look, it's Andre!

Typical Questions

It wasn't you?
Does anyone have a rubber band?
Oh, Tracey, what's wrong with us?
Can I join you?
How are you?
Are you sure?
Do you want one of mine?
What can I say?
Why are you telling me this?
What is that supposed to mean?!

TABLE 2

Greatest Hits! 25 Funniest Prelaugh Comments

Humorous Statements

He didn't realize he was sitting in dog shit until he put his hand down to get up.
When they asked John, he said he wanted to grow up to be a bird.
Look at that hunk of burning love.
He has a job holding back skin in the operating room.
Poor boy looks just like his father.
He tried to blow his nose but he missed.
You smell like you had a good workout.
She even makes my tongue hard!
I never eat anything that moves. (Reference to dormitory food)
Now you know what happened to Jimmy Hoffa. (Reference to dormitory food)
That's because you're a male!
I'd pay a hundred dollars to wade through her shit! (Expression of endearment)
She's working on a Ph.D. in horizontal folk dancing.
You don't have to drink, just buy us drinks.

She's got a sex disorder—she doesn't like sex.

You just farted!

Humorous Questions

Was that before or after I took my clothes off?

Do you date within your species?

Did you find that in your nose? (Reference to dormitory food)

Are you working here or just trying to look busy?

Why would you go water skiing if you don't know how to swim?

What did you do to your hair?!

Did he discuss anything during his last lecture? (Student query about a missed college lecture—one of mine!)

Is that considered clothing or shelter?

Are you going to wear that?!

The conversations we recorded were monitored long enough to establish that the low level of humorless prelaugh comments reported here was not an artifact of sampling only the out-of-context punch line of a joke or climax of a funny story. The low rate of joke-triggered laughter also was not the result of neglecting sight gags, comic gestures, or other visual cues, because plenty of laughter is present in telephone conversations, a purely auditory mode of communication. The telephone is a good, low-cost filter that passes auditory cues while blocking all visual ones.

Most prelaugh dialogue is like that of an interminable television situation comedy scripted by an extremely ungifted writer. When we hear peals of laughter at social gatherings, we are not experiencing the result of a furious rate of joke telling. The next time you are around laughing people, examine for yourself the general witlessness of prelaugh comments.

The discovery that *most laughter is not a response to jokes or other formal attempts at humor* forces a reevaluation of what laughter signals, when we do it, what it means, and how we should study it. Humor (jokes, pranks, tricks, gags, cartoons, etc.), taps only recently evolved cognitive and linguistic stimuli for laughter (e.g., incongruity) of the sort that concerned Kant and Schopenhauer and was described in Chapter 2. Although valid in their own domain, humor-based approaches are of limited relevance in understanding most laughter.

Compare, for example, the socially impoverished, narrow, and verbally oriented scenario of stand-up comedy, the prototype of much laugh research and philosophical inquiry, with the richness and complexity of everyday conversational laughter.

1. Stand-up comedy is based on joke telling, in contrast to the mundane, nonjokey, prelaugh comments of everyday life.

2. The comedian, the designated joke-teller in stand-up comedy, is physically and socially distant from the audience, in contrast to the intimate contact and interaction during normal conversational laughter. (Comedians attempt to close this gap and develop a relationship with the audience with such stories as "Have you ever taken your car to a garage for repair and the mechanic says . . . " etc.)

3. Comedians typically talk but don't laugh, and their audience laughs but doesn't talk, an unnatural representation of the social setting of conversational laughter, in which the speaker laughs most of all, and periodically trades roles with the audience.

Laughing Relationships

"Ha-ha-ha" is not broadcast into the void like the message of an animal calling, "This is my territory" or "I'm available for mating." Laughter, like speech, is a vocal signal that we seldom send unless there is an audience. Danish philosopher Søren Kierkegaard noted the rarity of solitary laughter when he inquired of a friend, "Answer me honestly . . . do you really laugh when you are alone?" Kierkegaard concluded that you have to be "a little more than queer" if you do. Indeed, laughter is the quintessential human social signal. Laughter is about relationships.

To learn more about the social setting essential for laughter, 72 undergraduate student volunteers from my classes recorded their own laughter, its time of occurrence, and its social circumstance in small pocket-sized notebooks (laugh-logs). Smiling and talking were also recorded to provide contrasts with laughter and each other. Laughing and talking are principally auditory signals, functioning in light or darkness and around obstructions. Smiling, in contrast, is a visual signal requiring line-of-sight visual contact between the recipient and the illuminated face of the sender. Talking was studied because its role in communication is unquestioned. The approach to talking taken here is a bit unusual because no attention was paid to what was said—subjects simply recorded when talking occurred. Singing "Louie, Louie" was weighted equally with the recitation of the Gettysburg Address.

A decision had to be made about how to treat media. For example, are you really "alone" when sitting in your living room watching television, listening to the radio, or reading a book? Probably not—you respond to media as a source of vicarious social stimulation. Your feelings of fear, loathing, lust, love, and aggression produced by media is not a response to an arbitrary pattern of light, sound, or imagery, but the product of relationships you have formed with the characters and events portrayed. When experiencing vicariously the trials of Scarlet and Rhett, you really do give a damn. To control for these confounding influences, only media-free cases of solitary laughter were considered.

The dedicated logbook keepers revealed that laughter, smiling, and talking were infrequent immediately before bedtime and after waking. After all, we seldom have much of an audience at these times, and when we do, we are not enthusiastic communicators. Among *solitary* subjects, talking was by far the most common early morning activity, followed by smiling, with laughing a distant third. Singing, rehearsing upcoming conversations, studying, cursing, and "thinking out loud" were parts of these morning soliloquies. The monologues were sometimes punctuated with smiles, but seldom laughs.

The *sociality* of laughing was striking. (Sociality refers to the ratio of social to solitary performance of a behavior.) My logbook keepers laughed about *30 times* more when they were around others than when they were alone—laughter almost disappeared among solitary subjects not exposed to media stimulation. In the social sciences, where many effects are tiny and revealed only through statistical analyses of huge samples, it's gratifying to find a result of this magnitude. Laughter's marked sociality reflects its evolutionary roots in tickle and rough-and-tumble play, activities requiring a partner (Chapters 5, 6). In contrast to the awesome sociality of laughter, people smiled over six times more and talked over four times more in social than in solitary situations.

Although we probably laugh or smile more when we are happy than sad, these acts are performed primarily in response to face-to-face encounters with others, our audience. This dependence on social context means that, contrary to popular opinion, laughter and associated facial behavior are unreliable mood meters—after all, would you announce "I'm a very happy person," when entering a vacant room or striding down a sidewalk alone?

Until recently, bowlers have been known mostly for beer, big hair, fancy shirts, and nondesigner shoes. But psychologists Robert Kraut and Robert Johnston changed all of that when they ventured into Ide's Bowling Lanes in Ithaca, New York, to observe smiling. In the most notable psychological research ever conducted at a bowling alley, they observed that bowlers often smiled during social interactions, but not necessarily after receiving good scores (strikes or spares). Furthermore, the bowlers rarely smiled while facing the pins, but often smiled when facing their friends. As in my logbook study, there was a strong association between smiling and social motivation and an erratic association with emotional experience.

Smiles are also flashed as social displays by athletes at the other end of the fitness spectrum from Ithaca's bowlers. José Miguel

Fernandez-Dols found Olympic gold medal winners at the Barcelona games to smile fleetingly when receiving their medals but only intermittently at other times during the presentation ceremony. The smiles were coupled with the face-to-face encounter with the presenter although the unquestionable joy of the moment, the culmination of a lifelong quest, was presumably stable during the entire presentation. The use of the smile as a social display develops early in life. Babies at play tend not to smile until they look around and make eye contact with their mother, observed psychologist Susan Jones.

Eye contact between friends (the people you laugh with) also facilitates laughter, a discovery I made while interviewing strollers on Baltimore sidewalks. When I encountered oncoming pedestrians (I'm sometimes trailed by a video cameraman) and informed them that "I'm conducting a survey of laughter," they typically shifted their gaze to each other and laughed. When I asked why they were laughing at each other but not at me, they often said, "You aren't funny," a rationalization of their action. People's nonverbal behavior tells a less ambiguous story. Laughter is a social act involving members of their group (their companions), and eye contact is an important link in this social pas de deux. I was the uninvited and "unfunny" outsider. ("Funny" and "unfunny" are simply ways of saying that you laughed at or did not laugh at somebody or something. "Funny" is not an adequate explanation of laughter.)

The social circumstances that most favor laughing and smiling are similar to those that favor talking. Talking may be more akin to laughing, smiling, and other nonverbal social signals than is often appreciated. For example, I have suggested that small talk may have evolved to facilitate or maintain social bonds among our tribal ancestors, a role independent of linguistic content and similar to that served by mutual grooming among members of contemporary primate troops. Robin Dunbar developed this proposition in his entertaining book, *Grooming, Gossip, and the Evolution of Language.* In her study of gender differences in conversation patterns, Deborah Tannen concurs that "small talk serves a big purpose," being "crucial to maintain a sense of camaraderie when there is nothing special to say." This social bonding function is a property of the "phatic" speech described by Malinowski "in which the ties of union are created by a mere exchange of words." In this context, the act of speaking is more important than what is said.

Laughter plays a somewhat similar, nonlinguistic role in social bonding, solidifying friendships and pulling people into the fold. You can define "friends" and "group members" as those with whom you laugh. But laughter has a darker side, as when group members coordinate their laughter to jeer and exclude outsiders. *Ridicule,* a fine French film exploring this theme, begins with the apocryphal quotation "In this country, vices are without consequence, but ridicule can kill" (Duke of Guines). The film deftly instructs us on how wit and laughter were used as currency and weapon in the effete, socially competitive court of Louis XVI. But laughter has long been instrumental in the casting out of misfits, sometimes with dire consequences. Throughout the ages, cripples, mental defectives, and court fools have been injured and perhaps even killed in a crescendo of teasing, laughter, and violence. Laughter scorns the victims and bonds and feeds the wrath of aggressors. On a more massive scale, dark laughter has sometimes accompanied the looting, killing, and raping that are among the traditional fruits of war.

We are still burdened with such savage vestiges of our primate heritage. Recent news reports confirm that mob violence, massacres, and butchery throughout the world are sometimes accompanied by laughter. During 1999, laughter has been reported during ethnic violence in Indonesia and Kosovo, and in a high school massacre in Littleton, Colorado. According to Aron Cohn, a survivor of the Littleton school shooting, the two male killers "laughed. They were just hooting and hollering. Having the time of their life" ("Death Goes to School with Cold, Evil Laughter," *Denver Rocky Mountain News,* 21 April 1999).

One of the best theatrical illustrations of the two sides of laughter appears in the film *Goodfellas,* in which the volatile character played by Joe Pesci at times "laughs with" and "laughs at" his mob buddies and outsiders, sometimes with deadly effect. Consider especially the notorious "Do you think I'm funny" nightclub scene early in the film.

What is the nature of a *laughing relationship*—the association necessary for a stimulus object, organism, or person to be considered funny (i.e., trigger laughter)? Most of us can envision a social relationship with a pet dog or cat, honorary members of our social world. We play with these creatures and may even laugh at them—the more humanlike their behavior, the funnier we perceive them. But can an inanimate object be funny? Here we revisit an issue developed by philosopher Henri Bergson (Chapter 2)—that mechanical things are funny in proportion to the degree in which they resemble humans (i.e., puppets are funnier than auto transmissions). Can you imagine a truly funny device, a creation that I will term a Bergson machine?

Let's consider the fruit test, a botanical twist on the Bergsonian proposition. Can a fruit be funny? What's the funniest fruit you can think of? Imagine the firm roundness of a grape, the plump purple majesty of an eggplant, or the blood red juice of a pomegranate. Although you can do funny things with fruit, by itself, fruit just isn't funny. For something to be funny it must be associated with the actions of people, not objects. The lowly banana peel has earned classic status in comic annals—but only when someone slips on it.

The film industry has important lessons to teach us about generating emotional reactions to inanimate objects—it's a multibillion-dollar laboratory for the study of human behavior that produces data in the form of box office receipts. No scientist has access to such wonderful resources. In the specialized domain of laughter and positive affect, we are far from building a Bergson machine that could win an audition for stand-up comedy. However, the *Star Wars* films have provided promising robotic contestants in those personable, intergalactic swashbucklers C3PO and R2D2. But even here, one member of this squabbling duo, C3PO, is humanoid, while R2D2 resembles a mobile, high-tech fireplug. Their droid humor may fail if it involved only a pair of R2D2s.

Is Laughter Consciously Controlled?

Do we choose to laugh? Do we decide to say "ha-ha-ha" as purposefully as we would select a word in speech? This is one of the most important and neglected questions about laughter; the validity of over two millennia of thinking about laughter hangs in the balance. Many philosophical and social scientific analyses bear the tacit assumption of intentionality and conscious control, and we saw its marks in the common sense rationalizations of why we laugh in Chapter 2. It is understandable that we seek answers in the familiar, but common sense does not serve us well in the unconscious realm of laughter. The narrow beam of consciousness cannot illuminate behavior that lies beyond its reach, and this beam is not only highly selective, it is turned off a surprising amount of the time. In this exploration of voluntary laughter, we will abandon the myth of human rationality and self-control and let simple behavioral demonstrations be our guide.

To learn about the voluntary control of laughter, I contacted Sam McCready, master actor and director in the University of Maryland Baltimore County's fine theater department. This seemed like a logical first step, in that actors are fellow experts in human behavior, their work having been shaped by the ruthless peer review of audiences for more than 2,000 years. At various times, Sam has been the Marquis de Sade, King Lear, Tartuffe, and any number of saints and sinners. Today's role involved his teaching an improvisational acting class, and he invited me to participate in their "laughing exercise." His students gathered in a large circle and one by one each attempted to laugh. Individually, their efforts were not impressive—most of their laughs sounded forced and artificial. They laughed more convincingly when they gathered in groups of two and four and laughed with and at each other—their difficulty in laughing on cue became a legitimate trigger of involuntary laughter. In their indirect way, the struggling novice actors were announcing that *laughter is under weak conscious control.*

Note three simple but informative demonstrations of the weak conscious control of laughter conceived during the student exercises.

1. *Ask a friend to laugh.* Most will respond immediately with a burst of genuine laughter (the *command effect*), after which about half of them will announce "I can't laugh on command," or make some equivalent statement. Your friends' observations that they can't laugh voluntarily are accurate—their subsequent efforts to laugh on command will be forced or futile. (This lack of voluntary control does not hold for all expressions of positive affect. When asked to smile, people will comply immediately and will never comment that they can't smile on command.)

2. *Ask your now laughless friends to say "ha-ha-ha."* They will comply immediately with a hearty "ha-ha-ha," an approximation of the sounds of genuine laughter. The difficulty of laughing voluntarily in the first demonstration was not, therefore, due to an inability to form the sounds of laughter—it was an inability to access the neurological control mechanism for spontaneous laughter. The present demonstration also indicates that normal, spontaneous, laughing is not the matter of speaking "ha-ha-ha."

3. *With a stopwatch in hand, ask someone to "Laugh when I say 'now!'" and measure the latency to the first laugh.* The reaction time may be many seconds, if the person can laugh at all. Contrast this long reaction time with the almost immediate response to the command "Say 'ha-ha-ha' when I say 'now!'" Long reaction times are also associated with requests for sobbing or crying, other emotionally laden vocalizations. We cannot easily and deliberately access and activate the brain's mechanisms for affective expression. ("Method actors" of the Stanislavski school circumvent the limited voluntary access to emotional expression by imagining a situation in which the desired response is associated.)

The indignant "Ha!" is an example of voluntary, consciously controlled laughter in speech. Another case is laughspeak, a hybrid of laughing speech that communicates emotional tone and shows more flexibility than classical ha-ha type laughs. Laughspeak is especially notable in talk-show hosts, public relations professionals, and others who attempt to defuse a delicate inquiry by posing it in a laughing voice. It is also common when there is a power imbalance between conversants, as when one person is trying to ingratiate themselves to another. A striking case of this occurred with a female host of a local classical music program who produced a lot of laughspeak and giggling in the presence of prominent, usually male guests, such as famous soloists and conductors, but not in the presence of lesser lights. This effect was highly reliable, and her laughter was usually unreciprocated, in line with the previous observations of gender differences and power relationships. The power dynamics of laughter and laughspeak are obvious and quite entertaining once you develop an ear for them; they are also penetrating and uncensored because of the unconscious nature of most laugh production and the limited awareness people have about their own laughter.

YOU LAUGH—YOU LOSE read the poster advertising a comedy game show to take place one evening in the University of Maryland Baltimore County's student union pub. "Survive one minute of comedy without laughing and spin the comedy cash wheel to win up to $200." The producers of this show were providing me with a nice little experiment about the control of laughter, and all I had to do was show up. This naturalistic experiment complements the just described efforts to produce laughter voluntarily—the task here is to *inhibit,* not produce laughter. Anecdotal evidence suggests that we are better at intentionally inhibiting than producing laughter, but sometimes laughter is hard to contain, as during laughing fits and contagious outbursts (Chapter 7).

Once the show was under way, the actual rules of the game were clarified—a contestant would lose if he laughed *or* smiled (!) during a one-minute comedy barrage from one of three professional comedians. These comics were playing hardball—smiles are more readily activated than laughs, and they are harder to suppress. The large audience groaned in disapproval. As it turned out, the comics were the ones having to sweat. Only two of the six contestants were eliminated; their smiles were so subtle that they were invisible to the booing audience. In the end, no contestant laughed. The comics were funny to the audience but they were clearly defeated in this confrontation with human nature. The audience members were entertained and treated to a demonstration they didn't anticipate—that *it is easier to inhibit laughter than it is to inhibit smiling.* The expressionless poker face is harder to maintain than laughlessness.

Smiling is a more nuanced, subtle medium than the ballistic blast of laughter, and its threshold for activation is much lower. Thus, smiling is a relatively "leaky," difficult to suppress channel of affective communication. The production of smiles is, however, under more voluntary control than laughter—if you ask someone to smile, they can comply immediately. Paul Ekman and Wallase Friesen have shown that such voluntary (false) smiles are subtly different from their spontaneous (felt) counterparts, but such "false" smiles are true in the way that counts most, in being an effective, socially potent means of communication. The voluntary smile is an important evolutionary adaptation that provides increased flexibility and conscious control over facial behavior.

Neurological accidents and disease are unfortunate "experiments of nature" that further distinguish spontaneous and voluntarily emotional displays. Consider central facial paralysis, which prevents patients from voluntarily moving the left or right side of their face to smile or produce any other expression. Such left or right "hemiparalysis" is produced by damage to higher brain motoneurons, leaving intact the neural pathway between the brain and face. When asked to grin, these patients produce crooked smiles—only one side of their face responds. However, they produce a normal, symmetrical smile if tickled or amused by a joke—the ongoing social stimuli activate intact neuronal pathways that are beyond conscious control. Here we glimpse the otherwise invisible hand of the ancient neurological puppeteer that controls spontaneous laughter and smiling.

The symptoms of Parkinson disease offer an informative contrast to central facial paralysis—voluntary facial movement is spared, but spontaneous facial movement is impaired. Parkinson disease is a disorder of the motor system that is associated with damage to the dopamine neurotransmitter pathway in the brain. During spontaneous social interactions, many Parkinson patients exhibit a bilaterally symmetrical, masklike face, but are able to smile normally on command.

A study of "split-brained" neurosurgery patient P. S. by Michael Gazzaniga and Joseph LeDoux offers a fitting conclusion to this chapter by reinforcing its lesson about the limits of self-reporting and nonempirical approaches to understanding laughter. P. S. had the main connection between his cerebral hemispheres (the corpus callosum) severed to control the spread of intractable epilepsy from one side of his brain to the other. The bilateral flow of information was also curtailed—one side of his brain did not know what the other side was doing. As a result of having bilateral representation of language comprehension, P. S. could respond to verbal commands presented to either hemisphere, but he could describe verbally only left hemisphere stimuli. (His right hemisphere was mute.) Although his talking left hemisphere could not access the knowledge of his surgically isolated right hemisphere, it would gamely attribute cause to right hemisphere action. For example, when his right hemisphere was ordered to respond to the printed command "laugh," he chuckled. When asked to explain the laughter, his ignorant but talkative left hemisphere responded "Oh, you guys are really something," an effort to rationalize his apparent bemusement.

We do not fare much better than patient P. S. when explaining why we laugh. Language and logic fail us when we venture into the nonverbal and nonrational realm. We must let laughter speak for itself through objective behavioral measurements and descriptions, and not impose our designs on its cryptic message.

FOUR

Cracking the Laugh Code

From Sound Lab to Opera Studio

hat *is* the sound of laughter that so affects our lives? From infancy to old age, whether Manhattanite or Zulu, we produce, hear, seek and avoid this yet undefined utterance in the universal human vocabulary. Guffaws, chuckles, and giggles are not arbitrary sounds, but types of innate acoustic code that have priority access to our brains and trigger behavior. Barriers between cultures and races fall away when we use this ancient signal, a sound more like an animal call or cry than human speech. But what, really, is laughter? In this chapter we get down to the nitty-gritty of describing laughter—of cracking the laugh code—an endeavor central to this book. Success in identifying the unique acoustic signature of laughter brings rewards beyond merely understanding the action and sound of laughter.

"Ha" is a modest vocal quantum that lacks the daunting complexity that gives language its grandeur. Laughter's stark, skeletal structure simplifies our task of describing it and makes laughter a powerful probe into neurobehavioral processes. As a "simple system," laughter offers a manageable approach to difficult issues of speech production, perception and evolution, just as fruit flies serve in genetics, tissue culture in cell biology, or nematodes in the study of neural development. The elemental, stereotyped laugh, for example, is a good starting point for understanding how we produce the myriad sounds of speech. And in the perceptual domain, some aspect of "ha-ha-ha" is the acoustic trigger of the contagious laugh response, the strongest evidence in humans of a neurological detector for a specific vocalization (Chapter 7). The following chapters build upon the description of laughter, first examining the evolutionary heritage of laughter in humans and chimpanzees, and later considering the mechanism that unleashes the noisy burst of contagious laughter.

The effort to crack the laugh code involves a two-pronged attack, one modern and technical, the other ancient and aesthetic. The first approach brings machine acoustic analysis to bear on recorded laughter. The second mines opera scores and recordings for clues about laughter's sonic structure. The hilarity portrayed on the operatic stage is cued by musical notation, the most rigorous, commonly available system of sound description of the pretechnological era. Here the musical shorthand of opera is converted to real time and treated as data. This analysis matches the analytic skills of the musical greats against the sound spectrograph, their modern electronic challenger.

The Acoustic Analysis of Laughter

The first step in sonic analysis is to obtain clean high-fidelity recordings of laughter for study, a more formidable task than it seems. Once again we encounter a variant of the pesky "tiger soup" problem of Chapter 3. Surreptitiously recording people's conversations runs afoul of right-to-privacy laws, and recordings made during comedy performances are usually full of extraneous noise and audience laughter. The most successful means of collecting laugh samples was to walk up to people in public places, audio tape recorder in hand, and announce "I'm studying laughter. Will you laugh for me?" This request usually was followed by a burst of hearty and genuine laughter (the command effect), especially when I adopted a slightly goofy demeanor. The highest laugh yield was achieved when solicited from a group of friends. When one member of the group is asked to laugh, the others will laugh at the plight of their pal, triggering rounds of give-and-take laughter that may last for many seconds.

With a hard-won sample of 51 laugh recordings, I retreated to the sound lab to learn what gives the laughness to laughter. I was joined in this task by my talented undergraduate assistant Yvonne Yong. We used a sound analyzer to generate detailed graphic and numerical descriptions of sound frequency and amplitude. One of the most useful outputs of this device is the sound spectrum, sometimes referred to as a "soundprint," a snapshot of a sound.

Using the sound spectrograph, we quickly discovered the distinct acoustic signature of laughter. You can visualize laughter as a series of evenly spaced sonic beads on a string. Each "bead" corresponds to a short, vowellike *laugh-note* (syllable) (i.e., "ha," "ho," "he," as transcribed in English) that has a duration (diameter) of about $\frac{1}{15}$ second (75 milliseconds). The beads are spaced at regular intervals (onset to onset) of about $\frac{1}{5}$ second (210 milliseconds). Laughs typically proceed with a decrescendo, a gradual reduction in loudness as the laugh progresses (Figure 4.1).

Laughter is not defined by a specific vowel sound, because there are many variants, such as "ha-ha-ha," or "ho-ho-ho," or "he-he-he." However, notes (beads) in a given laugh tend to have a homogeneous structure—you may hear "ha-ha-ha-ha," or "ho-ho-ho-ho," but not "ha-ho-ha-ho" laughs. There are biological constraints against producing such mixed-note laughs. Try to stimulate a vigorous "ha-ho-ha-ho" laugh—it will be difficult and it will feel and sound quite unnatural, if you can accomplish it at all. "Ho-he-ho-he," "he-ha-he-ha," and other mixed-note laughs are equally challenging. Our vocal appara-

Common Laugh Variants

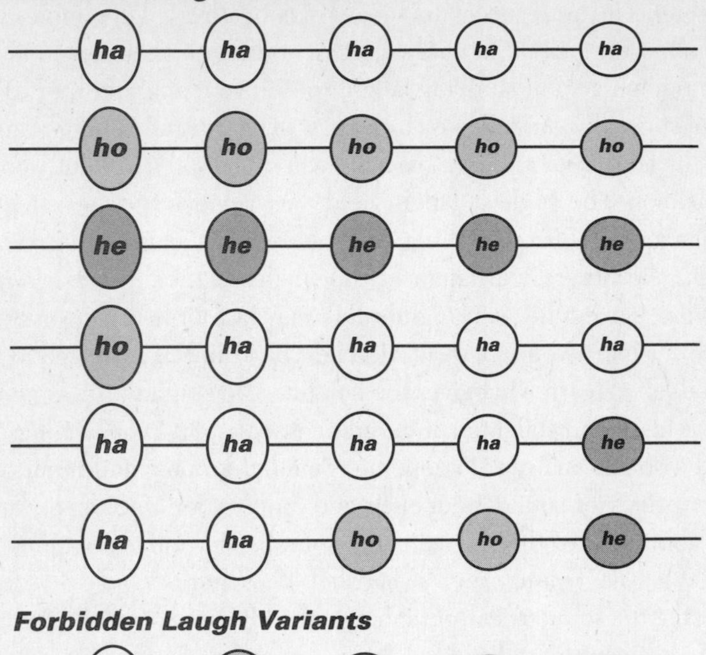

Forbidden Laugh Variants

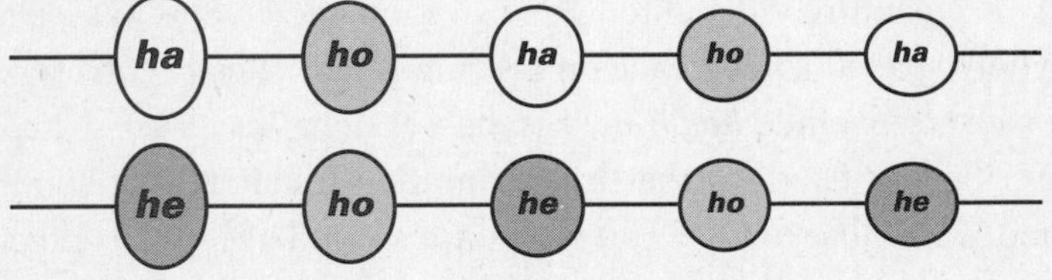

Figure 4.1 Variations of a five-note laugh. Laughter can be visualized as a series of beads on a string, with each "bead" having a duration of about 1/15 second and repeating every 1/5 second. Some laugh variants are common and easy to mimic, while "forbidden" types do not occur naturally and are difficult to produce; a sample of each type is shown. The gradual decrease in note amplitude is due to the loss of air needed to produce notes late in a sequence.

tus discourages such vocal gymnastics, though you can also easily switch note types in mid-laugh, as in "ha-ha-ho-ho," or the more exotic "ha-ha-ho-ho-he-he." Variations in laugh-notes, when they occur, most often involve the first or last note in a sequence. Thus, "cha-ha-

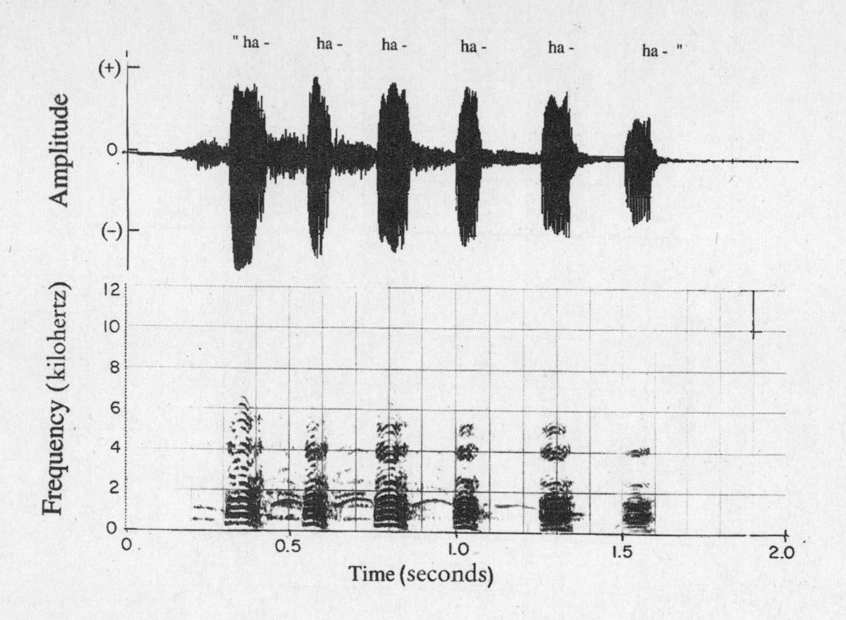

Figure 4.2 The unprocessed wave form (top) and its frequency spectrum (bottom) of a six-note *ha-ha* laugh from an adult male. The spectrum of each laugh-note is composed of a stack of horizontal stripes (a harmonic series) in which each stripe is a multiple of a low fundamental frequency, the bottom stripe of each stack. In both the waveform and the spectrum, a weak signal occurs before and between each laugh-note; when not masked by adjacent laugh-notes, it sounds like a sigh. (Adapted from Provine, 1996)

ha" or "ha-ha-ho" laughs are possible alternatives, and are almost as easy to produce as the homogeneous "ha-ha-ha" or "ho-ho-ho."

Let's now consider the fine structure of laughter as revealed in the sound spectra. The most striking feature of the laugh-note spectrum is the vertical stacks of horizontal stripes that are immediately recognizable as a harmonic series, meaning that each stripe is a multiple of a low fundamental frequency, the bottom stripe. Thus, the harmonics of a 200 Hz tone (the fundamental) would be 400 Hz (the first harmonic), 600 Hz (the second harmonic), etc. (Figure 4.2). Women's laughter usually has a higher fundamental frequency

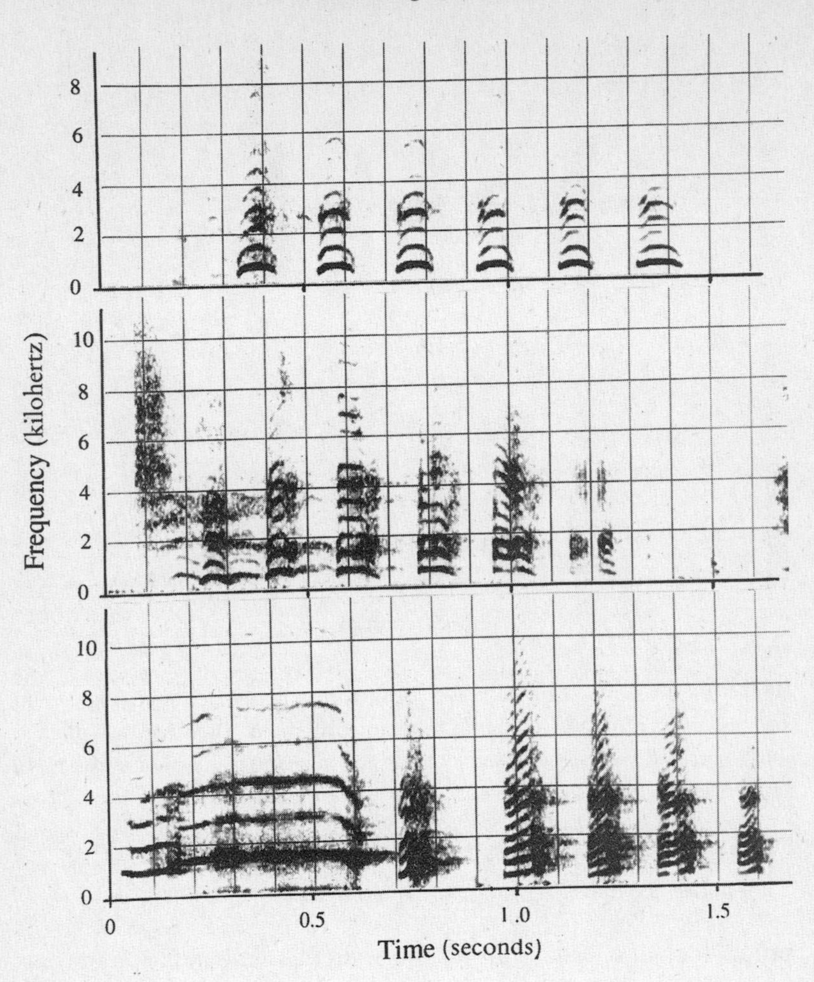

Figure 4.3 Laugh variants of adult females. At top is a *he-he* laugh with the widely spaced harmonics associated with its high frequency. The second is a *cha-ha-ha* laugh (not all laughs begin with /h/). The bottom example is a high-pitched squeal (*eeee!*) with strong harmonic structure that turns into a *ha-ha* laugh. (The gleeful squeal was a response to the threat of being tickled by this young woman's male friend.) Although the present focus is upon the central tendency about which laughter varies, the snorts and unvoiced laugh sounds sometimes made by both males and females lack the clean note and harmonic structure shown here.

(~502 Hz) than men's laughter (~276 Hz), a finding consistent with women's higher-pitched voices. The harmonics (stripes) of women (and children) are more widely spaced than those of men because each is a multiple of a higher fundamental frequency (Figure 4.3).

An experiment with digitally modified laughs revealed that a weak "sigh" precedes and follows all laugh-notes. This work was conducted with the assistance of Mike Cerri, director of the University of Maryland Baltimore County's recording studio. Using a digital sound-editing and analysis system, Mike removed the laugh-notes from a recorded laugh and closed the gaps previously occupied by the notes. (Imagine stripping the laugh-beads, leaving a naked string.) When we listened to the edited laughs, sans notes, we heard a long, unvoiced, breathy aspiration—a sigh. In normal experience, this sighing goes unnoticed because it's masked by the loud blasts of the adjacent laugh-notes. The negligible contribution of the prenote aspirations to the sound of laughter was demonstrated by editing out the sounds surrounding laugh-notes, leaving only naked laugh-notes (beads) separated by intervals of silence. The resulting laughter sounded normal. Thus, the critical information for the identification of laughter is borne by the notes and by the time-interval separating them. The breathy sound that initiates and follows each laugh-note serves no communication function and is probably a by-product of laugh-note production.

Earlier we saw that the vocal mechanism conspires against the production of mixed-note laughs (e.g., "ha-ho-ha-ho"). Now let's consider vocal constraints in the time domain.

It is difficult to laugh with abnormally long note durations, such as "haaa-haaa-haaa," or abnormally short, pizzicatolike durations (much less than the typical 1/15 second note length). As with note-type ("ha," "ho," etc.), variations most often involve the beginning or end of a laugh episode. "Ha-ha-ha-haaaa" or "haaa-ha-ha-ha" are likely variants, but not "ha-haaa-ha-haaa," etc.

Laughs with abnormally long or short internote intervals are other forbidden options. The best evidence comes from a self-demonstration. Try to laugh with a long internote interval, such as "ha———ha———ha," instead of the usual "ha-ha-ha." As with the natural rhythms of walking or running, there are only so many ways to laugh.

The string of laugh-notes (beads) looks basically the same whether you scan it from left to right or from right to left along the horizontal time axis. This temporal symmetry suggests that an audio recording of laughter should sound similar whether played in the usual forward direction or in reverse. Indeed, "ha-ha-ha" played backward on a tape recorder still sounds rather like "ha-ha-ha"—it certainly maintains its pulsative character and fares far better than speech, which becomes unintelligible when reversed.

The unvoiced /h/ sound associated with the beginning of the vowellike laugh notes ("ha" as transcribed in English) may arise in part within our auditory system as the product of the rapid onset of the note—otherwise, it would not still sound like "ha" when reversed. Additional indirect evidence of laugh-note symmetry is found in the comparison of musical laugh notation in different languages. Borrowing from the opera (below), we learn, for example, that in Italian, laughter is notated as "ah" in contrast to "ha" in a parallel English translation—the Italian language has no equivalent /h/ sound. Although I have not put the proposition to formal test, both English and Italian speakers probably differ more in the *perception* than the production of the species-typical laugh sound. The opera literature was the source of other laugh esoterica. We learn, for example, that laughter is "ah" in French, "hi," "ha," "ho," or "ei" in German, "xa" in Russian, and "hej" in Czech, and yet the resulting vocalization is essentially the same.

One aspect of laughter that is certainly not temporally symmetrical is loudness. It's startling to hear reversed-laughter proceeding to a dramatic and bizarre-sounding crescendo, sounding a bit like a Woody Woodpecker laugh. In our lifelong casual experience with normal laughter, we never notice the corresponding decrescendo—it takes the prompt of hearing laughter played backwards to direct our attention to this phenomenon. The decrescendo of normal laughter is probably caused by our running out of air for laugh-notes late in a sequence.

One of the most striking features of laughter is its stereotypy—most people laugh in similar (but not identical) fashion. After all, if there was not some invariance in laughter, we would not recognize that people were laughing, and laughter would be useless as a social signal. As we have seen, our nervous system and vocal track enforce this stereotypy—we can't laugh in arbitrary ways even if we try.

Although laughter is stereotyped, it's not totally inflexible. People laugh in different ways at different times, expressing social, grammatical, and emotional nuance. However, our machine analysis would be hard-pressed to detect the subtle sonic distinctions between "sly," "wicked," "ironic," "gay," "hearty," and "sardonic" laughter, especially when essential cues may be contextual. Other shadings may be specific to the native language of the laugher—such differences may involve note intonation, inflection, or pitch, rather than the neuromuscular process that blasts out the universal rhythm of "ha-ha-ha." Individual differences also exist—a person's laughter is probably as distinctive as his speaking voice.

After considering laughter's central tendencies, those common features that we all share, I began to crave some exotica, those special people who are graced with bizarre-sounding laughs. I am not referring to individuals who laugh too much or inappropriately, but those whose vocal trumpeting pushes the envelope of weirdness. These are the people who can make your day—their honks, snorts, and brays are guaranteed to trigger a storm of contagious laughter. The possessors of such vocal gifts may, however, value their performance differently, fearing the mirthful eruption that may bring unwelcome attention, if not ridicule.

People with such unusual laughs can make a valuable scientific contribution—they establish the extremes of the distribution and may reveal how normal laughs are produced. For example, some individuals with long, honking/braying laughs neither voice nor chop the exhalation of laughter into notes—their laughter is one long breathy aspiration, as if the /h/ sound of the laugh-note is sustained without the vowellike "a." Other unexplored variants may involve sustained laugh notes (e.g., "haaaaaa"), unusually long, short, or irregular internote intervals, or disruptions of the punctuation mechanism that regulates the time-sharing of laughter and speech. (Pathological laughter is explored in Chapter 8.)

Odd laughs provide convenient behavioral markers for studies of heritability. They should be easier to recognize and trace across generations than the more complex and subtle shadings of speech. Sadly, biographers seldom chronicle their subject's giggles, honks, and chortles. Does the possessor of a really bizarre laugh manage to procreate? Are unusual laughs gender-specific? Do they run in families? Which, if any, parameters of laughter are heritable? Reports of laughter in congenitally deaf and blind children establish the innateness of the basic act, but leave open the possibility that aspects of laughter may be shaped by experience.

The laughter of identical twins is particularly informative because aspects of laughter may be "emergenic," that is, influenced by a configuration (rather than by a simple sum) of polymorphic genes. Emergenic traits do not run in families and may only be recognized as genetic if one studies monozygotic twins who share all of their genes and gene configurations. (Running speed in race horses is an emergenic trait, for example.)

In the absence of genetic studies of laughter, we must turn to limited, though provocative, anecdotal evidence. A genetic predisposition for laugh styles is suggested by the identical "giggle twins" who were separated at birth and not reunited until 40 years later when they participated in the famous Minnesota Study of Twins Reared Apart. Until they met each other, neither of these exceptionally happy ladies had known anyone who laughed as much as she did. Yet, both were reared by adoptive parents they described as undemonstrative and dour. These gleeful twins probably inherited some aspects of their laugh sound and pattern, readiness to laugh, and perhaps even taste in humor, but we are far from understanding which traits were inherited and how such genetic predispositons influence laughter and steer us down life's chancy path.

A Night at the Opera

Among the murder and mayhem, yearning, lusting, and dying in the musical melodrama of opera, singers are sometimes allowed an occasional chuckle, yuk, or titter. But are these musically notated laughs useful data? Can we tap the fabled musical powers of a Mozart to capture the essence of laughter? Does the composer of *Die Zauberflöte* provide insights into *Die Zauberlachen*? And what about Puccini, Purcell, Rossini, and Strauss?

A musical score is a set of instructions to an instrumentalist or vocalist about the duration, pitch, and shape of a sound to be produced. To the scientist, a musical score is a quantitative description of the distribution of behavior in time. If you know the time signature ($\frac{2}{4}$, $\frac{3}{4}$, $\frac{4}{4}$, etc.), metronome marking (quarter note = 60 beats per minute, etc.), note value (whole, half, quarter note, etc.), and pitch (E, G, B, etc.), you can represent and reproduce a sonic event with reasonable accuracy. This is not a novel insight; in the days before oscilloscopes and sound spectrographs, ornithologists used music notation to represent bird song. So how accurate is musical notation in describing the "human song" of laughter?

The Repertoire of Laughter

The concept of examining the arithmetic conversion of giggles into acoustic data seemed straightforward and fun. The problem was in finding laugh passages among the tens of thousands of pages of weighty opera scores. Over the course of several years I assembled a wide-ranging sample of 60 operas and vocal works that had laughter in their score, libretto, or recorded performance (see list at the end of this chapter).

What did I find in this compilation of operatic laughter? As the house lights dim and the velvet curtains part, settle into your pricey front-center seat and get ready for a wild ride. Much operatic laughter occurs in a decidedly grim setting, far from the lighthearted chirping in the spirit of Johann Strauss's laughing song, *Mein Herr Marquis.* French feminist critic Catherine Clément develops the thesis of *Opera, or the Undoing of Women,* but opera is responsible at various times for the "undoing" of almost everyone. As Bugs Bunny once said, "Well, what did'ya expect in an opera, a happy ending?"

Laughter merges with cannibalism in Stephen Sondheim's *Sweeney Todd,* where the inventive Mrs. Lovett recycles the by-product of Sweeney's homicidal tendencies in her meat pies ("A Little Priest"). Composer Harrison Birtwistle contributes *Punch and Judy,* in which Punch sings a most un-Brahmsian lullaby to a baby and laughs wickedly before chucking the baby into a fire. In Reimann's German language adaptation of *Lear,* we are treated to a reenactment of the most gruesome scene in Shakespeare ("Out vile jelly.") in which Cornwall gouges out noble Gloucester's eyes. The event is celebrated by the evil, hysterical laughter of Lear's daughter, Regan. Alas, a king's life is often difficult, as in Rimsky-Korsakov's *Le Coq d'Or,* where senile King Dodon unknowingly trades his queen to an astrologer for a magic chicken—a fair deal considering this most unpleasant queen. When the astrologer shows up to claim the queen, he is killed by the king, who is himself killed by the chicken. The queen then vanishes, leaving only her laughter. Other grim laugh-fare is found in Sommer's *Vocal Sympathy,* based on Dostoyevsky's *Crime and Punishment,* and Eaton's *Danton and Robespierre,* a musical adaptation of the execution of two martyrs of the French Revolution. If murder, institutionalized or free-lance, is not to your taste, laughter is also associated with betrayal in love (*Phantom of the Opera*—Webber, or *Der Dreigroschenoper*—Weill) and probate (*Gianni Schicchi*—Puccini). Not even the inanimate are spared. In the science fiction opera *Aniara* (Blomdahl), the evil Captain Chefone laughs before blaming a crew member for the death of the spaceship's grieving computer.

If you are not offended by opera's violence and body count, there is the frequent challenge of silliness, irrelevance, and improbability. We need not turn to PDQ Bach (*Hansel and Gretel and Ted and Alice: An Opera in One Unnatural Act*) or Anna Russell for parody—the original composers and librettists do fine on their own. In *The Nose,* a satire by Shostakovich, the runaway proboscis of a Major Kovalev, which takes human form, is finally captured and returned in its original shape to its owner, to the occasion of much rhythmic laughter. But we shouldn't worry too much about a bit of silliness in our entertainment—it may even be good for us. In *The Love for Three Oranges,* Sergei Prokofiev makes an operatic pitch for the medicinal powers of laughter (a theme of Chapter 9). *Oranges* features a melancholy and hypochondriac prince who is subjected to a dose of therapeutic laughter prescribed by the king.

Musical Notation of Laughter

Having surveyed the dramatic landscape of operatic laughter, let's now consider notated laughter as data. My collaborators were Lisa Griesman, undergraduate historian/psychologist/flutist and now physician, and my wife, pianist/musicologist Helen Weems. Our quantitative analysis of notated and sung laughter was based on a sample of 20 operas for which we had both scores and recorded performances. (These operas are designated by asterisks in the list at the end of this chapter.) We did not analyze all notated laughter—extremely long laugh-notes held as a whole note or fermata were omitted, as were laughs that were highly stylized, irregular, or sung as a cadenza. Such musical flourishes were clearly not efforts to represent the sound or cadence of naturally occurring laughter.

The starting point of our analysis was the natural laugh sound that composers and singers seek to approximate, the subject of the first part of this chapter. Recall that spectrographic and waveform analysis defined laughter as a series of laugh-notes ("ha," "ho," "he," etc.) that last about $\frac{1}{15}$ second and repeat every $\frac{1}{5}$ second or so. Dividing the onset-to-onset interval between laugh-notes by laugh-note duration indicates that the laugh-note occupies only about one-third the interval between adjacent notes—about two-thirds of the interval between notes is silence. In addition, most laughs, particularly lengthy ones, proceed with a decrescendo, a progressive decrease of loudness.

Our attempt at notating laughter in common $\frac{4}{4}$ time (4 beats to the measure) is shown in Figure 4.4. To indicate that laugh-notes occupy about one-third of the time between the onset of adjacent laugh-

Figure 4.4 Musical notations of laughter. Three of the many possible notation schemes are shown. Acoustic analyses indicate that laugh-notes occupy about ⅓ of the interval (onset-to-onset) between notes. The first two schemes acknowledge the silence between laugh-notes by using rests, with the triplet figure (three notes to the beat) being the most precise option. The naturally occuring decrescendo (decrease in loudness) as a laugh proceeds is indicated by the converging lines beneath the staff. The last and least sophisticated option has been taken by many composers—simply asking singers to laugh during a given interval.

notes, laugh-notes are written as the first note of an eighth-note triplet figure (3 notes to the beat), the second two-thirds of which are occupied by the silence of a quarter-note rest. The metronome marking is a very brisk 300 beats per minute (bpm), a tempo that yields a laugh-note duration of $\frac{1}{15}$ second, and an interval between notes of $\frac{1}{5}$ second (onset-to-onset), approximating normal laughter. (The tempo of 300 bpm, corresponding to an internote interval of $\frac{1}{5}$ second, was selected to simplify the presentation, instead of the empirically more precise 286 bpm corresponding to the 210 millisecond internote interval of natural laughter.) Although tempi around

300 bpm are nothing special for bebop lovers, it's beyond the upper range of most metronomes. The decrescendo marking (the long converging lines) beneath the musical notes indicates the gradual decrease in loudness of the four laugh-notes.

If the triplet notation seems overly complicated, the figure provides two alternative though slightly less precise options. By permitting laugh-notes to occupy one-half instead of one-third of the internote interval, the notes can be written as ordinary eighth notes separated by eighth rests. The eighth notes (two notes per beat) can be shortened if sung in a staccato (separated) manner (denoted by a dot below the notes), leaving more of the desired silence between laugh-notes. Simpler yet, laugh-notes can be scored as staccato quarter notes (one note per beat), leaving out the rests altogether. An option taken by some composers is to avoid notation and simply mark on the score the interval during which laughter should occur, establish the tempo, and let the artists sing it in rhythm as best they can. There is wisdom in this last alternative: It's difficult enough to laugh voluntarily, without the added burden of laughing in the proper pitch, duration, and rhythm.

Having considered the options for notating musical laughter, we now examine the approaches actually taken by composers. There is a bewildering array of notation schemes, ranging from an occasional thirty-second, half, and whole note to the more usual sixteenth, eighth, and quarter note. The pitch of laugh-notes was usually scored, but sometimes only a rhythmic marking was given. Of course, the musical fraction of note duration has no meaning apart from the context of the time signature (beats per measure) and tempo (beats per minute) in which they occur. For example, the shortest duration laugh-notes (sixteenth and thirty-second notes) were found in songs with the slowest tempi. Time signatures of the musical laugh passages were diverse, including cut-time, $^2/_2$, $^2/_4$, $^3/_4$, $^4/_4$, $^5/_4$, $^3/_8$, $^6/_8$, and $^{12}/_8$. Composers agree that laugh passages should have fast tempi, most often allegro (fast) and sometimes even presto or vivace (very fast). However, in regard to laughter, this usually isn't fast enough.

Lisa and I mathematically reduced the laughter from our 20 opera scores. Although this exercise was frustrating and fraught with musicological uncertainty, we found that most composers *overestimated* the duration of laugh-notes ("ha") and internote intervals ("ha-ha"). Whether due to artistic constraint (musical context) or perceptual error, composers usually made laughter slower (longer) than it really is. But there is a great range of laugh rates even for an individual composer, or within an opera or song. For example, in Mussorgsky's *Boris Godunov,* the tempi of four different laughter-bearing songs range from about 35 percent too slow to 50 percent too fast. (Mussorgsky's vocal music is full of laughter and he is more accurate than most composers in establishing its tempo.) Perhaps we shouldn't be too critical of scored laughter. Laughter carries a built-in trap for composers; although the cadence of songs must speed and slow with the demands of the dramatic and melodic moment, naturally occurring laughter is essentially constant in tempo. In operatic laughter, artistic flexibility collides with the rigidity of a vocal instinct. Bowing to the tyranny of the acoustically correct laughter would unreasonably constrain tempi and associated musical development.

Composers also fell short in acknowledging the silent interval separating adjacent "ha-ha"s, and laughter's natural decrescendo. With a few exceptions (e.g., Massenet in some laugh passages of *Manon*), composers did not notate rests between laugh-notes. The effect of rests was achieved more often by scoring staccato (separated) laugh-notes, the approach of Johann Strauss in "Mein Herr Marquis" from *Die Fledermaus,* and Henry Purcell in the "Ho-Ho-Ho" witches' choruses from *Dido and Aeneas.*

It's hardly surprising that opera scores don't reflect the decrescendo of conversational laughter, given that sung laughter usually occurs in the midst of a melody in full flight. Strauss's decrescendo during one laugh episode in "Mein Herr Marquis" is an exception, as is the laughter in the garden scene (Act I, Part II) of Verdi's *Falstaff.* The dynamic shading in Mussorgsky's *Boris Godunov* and Massenet's *Manon* is in the wrong direction, cuing a crescendo of laughter instead of the behaviorally appropriate decrescendo, and conveying emotional climax instead of sonic reality.

Opera is a musical play with sung dialogue; therefore, we can evaluate the linguistic as well as the melodic context of laughter. Do the exultant sopranos and impassioned tenors and baritones of these musical melodramas get their yuks in the right place? Only sometimes. "Mein Herr Marquis" and many other operatic songs violate the punctuation effect, the lawful tendency of both speaker and audience laughter to occur only at the end of statements, questions, and natural pauses in speech (Chapter 3). In "Marquis," for example, the perky soprano Adele sings "What a funny—ha-ha-ha—situation," not the more behaviorally accurate "What a funny situation—ha-ha-ha." Mozart gets it right in "E voi ridete" ("What Is So Funny?") from *Cosi Fan Tutte*, where Don Alfonso sings "I can't be serious! Ha . . . " etc. The famous laughter from *Pagliacci*'s tragic (and murderous) Canio occurs appropriately at phrase end, but is not notated—the score simply instructs the vocalist to "laugh bitterly" in the Italianate "Ah! Ah! Ah! Ah! Ah!"

For even the most rigorous scientist and musiciologist the analysis of opera scores is fraught with uncertainty. Even when tempi are clearly marked, for example, it's difficult to ascertain the performance standard in the composer's own era. The *New Harvard Dictionary of Music* acknowledges the problem of establishing tempo, offering the learned advice of Leopold Mozart, Wolfgang's father. The elder Mozart asserted that "even if a composer endeavors to explain more clearly the speed required . . . one has to deduce it from the piece itself, and this it is by which the true worth of a musician can be recognized without fail." If this advice was good enough for the Mozarts, it's good enough for us. We deferred to masters of musical interpretation and analysis, the conductors and performers whose musical opinions are memorialized in their recorded repertoire. An analysis of the time structure of laughter of 20 operas (two versions of each work, when available) was conducted by Lisa, a trained musician, who sat patiently, electronic metronome in hand, estimating the tempi and intervals between laugh-notes for each opera.

Consultation with the recorded repertoire further complicated our task of using musical methods to define laughter. Artistic freedom was rampant—sung laughter often deviated from what was written. During laughter, singers sometimes slowed the tempo or speeded it up by singing faster notes than scored (e.g., sixteenth notes when eighths were written). Some deviations from the score may have been made to ease the singing or imitate the naturalness of laughter. Sometimes laughs scored with different laugh-note durations (e.g., eighths and quarters) were performed with equal durations, or laughter notated as a single pitch was sung as a series of descending pitches.

The most flagrant variance from scores involved adding laughter where none was written, as traditional in the aria "E Scherzo od e Follia" from Verdi's *Un Ballo in Maschera.* (This performance practice may have begun with Caruso.) When it comes to emoting, opera singers, whether soprano, alto, tenor, baritone, or bass, would rather giveth than taketh away—we found no cases of scored laughter being omitted. Offenbach's "Quel Diner" ("The Tipsy Waltz") from *La Périchole* is another occasion of improvised chuckles, providing one of the most engaging laugh performances on record. In a vintage recording, soprano Claudia Novikova sounds a bit lightheaded and is very full of herself as she splashes silvery laughter about in a gay and coy performance.

At study's end, we face a sober reckoning: We learned a lot more about opera than about laughter. The empirical excursion into operatic laughter brought quirky enlightenment and welcome diversion from the scientific main event, but no breakthroughs. We discovered, as we have seen, that operatic laughter is usually scored and sung a bit too slow, and that decrescendo and punctuation effects are seldom recognized. With the exception of our own technologically guided notation schemes, the potential of musical notation to describe the sonic reality of laughter was usually unrealized. In opera, too many concessions are made to the tyranny of the musical moment; laughter is always in the service of song, seldom vice versa. Thus, the fabled ears of Mozart and colleagues never got a fair test in our analysis of their musical offerings. We are left with romantic musings about how these titans of the past would have fared in head-to-head competition against the sound spectrograph, musical John Henrys battling their modern electronic challenger.

OPERAS AND VOCAL WORKS WITH LAUGHTER IN SCORE, LIBRETTO, OR PERFORMANCE

(Asterisks denote inclusion in the quantitative analysis.)

*L'Allegro, il Penseroso ed il Moderato** (Handel)
Aniara (Blomdahl)
*Il Barbiere di Siviglia** (Rossini)
*La Bohème** (Puccini)
*Boris Godunov** (Mussorgsky)
Candide (Bernstein)
Cantata 212 (The "Peasant" Cantata) (J. S. Bach)
*Les Contes d'Hoffmann** (Offenbach)
Le Coq d'Or (Rimsky-Korsakov)
*Cosí fan Tutte** (Mozart)
*La damnation de Faust** (Berlioz)
Danton and Robespierre (Eaton)
Death in Venice (Britten)
The Devils of Loudon (Penderecki)
*Dido and Aeneas** (Purcell)
Doktor Faustus (Busoni)
Der Dreigroschenoper (Weill)
The Dream of Gerontius (Elgar)
The Golden Vanity (Britten)
Elektra (Richard Strauss)
*Falstaff** (Verdi)
Die Fledermaus (Johann Strauss)
Gianni Schicchi (Puccini)
Hansel and Gretel and Ted and Alice: An Opera in One Unnatural Act (PDQ Bach/Peter Schickele)

In these Delightful Pleasant Groves (Purcell)
Irmelin (Delius)
*Khovanshschina** (Mussorgsky)
Lear (Reimann)
The Love for Three Oranges (Prokofiev)
Lucy and the Count (Deak)
*Manon** (Massenet)
Mary, Queen of Scots (Musgrave)
Masquerade (Nielsen)
The Midsummer Marriage (Tippett)
*The Mikado** (Gilbert and Sullivan)
The Nose (Shostakovich)
*Le Nozze di Figaro** (Mozart)
*I Pagliacci** (Leoncavallo)
Peer Gynt (Egk—not Grieg)
La Périchole (Offenbach)
The Phantom of the Opera (Webber)
*The Pirates of Penzance** (Gilbert and Sullivan)
Punch and Judy (Birtwistle)
Rayok (Shostakovich)
*Rigoletto** (Verdi)
Das Rheingold (Wagner)
*Der Rosenkavalier** (Richard Strauss)
Siegfried (Wagner)
The Song of the Flea (Mussorgsky)
*The Spinning Room** (Kodaly)
The Starlight Express (Elgar)
Sweeney Todd (Sondheim)
Symphony 14 (Shostakovich)
Thames Pageant (Panufnik)
They All Laughed (Gershwin)
The Tragedy of Man (Ranki)
*Die Zauberflöte** (Mozart)

FIVE

Chimpanzee Laughter, Speech Evolution, and Paleohumorology

Do we reign alone as "the only creatures that laugh," as first suggested by Aristotle over 2,000 years ago? Centuries of folk wisdom mostly agree that laughter is a uniquely human behavioral trait. But is it true? Or are we once again to be knocked from our homocentric throne? To pursue this age-old question, I sought expert advice from our closest primate relative, the chimpanzee. While tickling chimpanzees in an attempt to stimulate laughter, I made two related discoveries—why chimps can't talk, and the locomotor transformation (bipedality) necessary for the evolution of human speech. As used in this study, laughter becomes a tool to study vocal evolution.

Contrary to popular belief, our hairy cousins the chimpanzees do produce a laughlike sound, as do gorillas, orangutans, and perhaps other primates. Charles Darwin, in his classic *The Expression of Emotions in Man and Animals,* reported that "If a young chimpanzee be tickled—the armpits are particularly sensitive to tickling, as in the case of our children—a more decided chuckling or laughing sound is uttered; though the laughter is sometimes noiseless." He also notes that "Young Orangs, when tickled, likewise grin and make a chuckling sound." Dian Fossey, in her book *Gorillas in the Mist,* adds that "Tickling between [gorillas] Coco and Pucker provoked many loud play chuckles." Given these and similar reports by primate vocalization experts including Peter Marler and Jane Goodall, need we pursue the question of ape laughter further?

The answer is a decisive yes and no—in science, as elsewhere in life, things are often not so simple as they first seem. What Darwin, Fossey, and others agree upon is that chimpanzees and other great apes produce a laughlike vocalization in circumstances (i.e., being tickled, playing rough-and-tumble) in which humans reliably laugh. These are useful observations, but short on essential details. Until we recently cracked the laugh code, too little was known about the structure of human laughter to make detailed comparisons with other species (Chapter 4). We simply did not know what to look at or what measurements to make. But once armed with knowledge of human laughter, it became apparent that chimpanzee and human laughter differed in important ways.

To study chimpanzee laughter, chimpanzees were needed, and one of the best places to find them in the United States is at the Yerkes Regional Primate Research Center in Atlanta, Georgia. This institution is the home of many of the ape world's most distinguished ambassadors to humankind. For this part of the study, I teamed up with Dr. Kim Bard, then director of the nursery at Yerkes and a foster mother to many young chimpanzees who are ill, injured, or receiving inadequate maternal care from their biological mothers. Although Kim and I observed the ongoing behavior of chimps of all ages, we focused on animals of less than one year of age because they are especially playful and laugh a lot. We also chose babies because we were going to tickle them: tickling is one of the most common, reliable, and naturally occurring triggers of chimp laughter. Chimps are remarkably strong, can become aggressively boisterous as they age, and may easily injure their human playmates. Whether tickling chimpanzees, the proverbial 500-pound gorilla, or a human playmate, it's important to have a consenting and friendly subject.

Our observations occurred within a fenced area just outside of the nursery building at Yerkes. The setting resembled a playground for a security-conscious preschool, complete with gym equipment. The tickler was Kim Bard or her assistant, Kathy Gardner. Tickling always occurred during playful interactions that ranged from light stroking, to play biting the shoulders and arms, to rolling around on the ground with an armful of exuberant young chimp. Both women were familiar to the chimps and were accepted as playmates.

When tickled by Kim or Kathy, the chimps produced a characteristic "play face" (mouth open, upper teeth covered, lower teeth exposed) and emitted a breathy pantlike sound that characterizes chimpanzee laughter.

Chimp laughter differs more from its human counterpart than is suggested by such previous descriptors as chuckling. Chimp laughter has graded variants, ranging from barely perceptible, labored breathing, to a more vigorous form in which a voiced, guttural exhalation overshadows the lower-amplitude inhalation. In cases of especially exuberant laughter, both exhalations and inhalations are voiced. The grunting sound ("ah grunting") noted by some investigators, best describes such high-amplitude chimp laughter. The few, sketchy descriptions of gorillas and orangutans suggest that they, too, make breathy, panting sounds when laughing, and exhibit a chimplike play face.

Audio recordings of chimpanzee laughter did not sound laughlike to students in two of my college classes. Almost no one hearing the tapes was able to identify the chimp sound as laughter (2 of 119), whereas almost everyone recognized adult male human laughter (117 of 119). Chimp laughter proved to be an auditory Rorschach test, triggering many associations in listeners. The most common description was "panting" (36 students)—most often believed to be that of a dog. Twelve students used other breathing-related descriptors (i.e., "asthma attack," "hyperventilation," "breathing problems,"

Figure 5.1 Play face of a young chimpanzee. The characteristic "play face" (mouth open, upper teeth covered, lower teeth exposed) accompanies pant-like chimpanzee laughter. (From Provine, 1996)

"breathing hard"). Some students noted only the animal suspected of making the sounds, including dog (10 students) or various nonhuman primates (ape, chimpanzee, monkey, or gorilla; 16 students). Other adventurous folks volunteered a variety of acts being performed during the vocalization, including "shivering" (3), "running" (2), and "masturbating" or "having sex" (5). A surprisingly large number of students (17) attributed the chimp sounds to nonbiological, mechanical acts, most commonly "sawing" (9), but also "scraping" (2), "erasing" (3), "brushing" (2), and "sanding" (1).

Despite the students' confirmation of my impression that chimp laughter sounds like panting, it's easy to understand why many sophisticated observers since Darwin have associated the chimp vocalization with very different-sounding human laughter. Anyone who has played with chimps has been impressed with their laugh-like behavior. Consider the powerful context cues of chimp laughter—it follows tickling, it is produced during the physical contact of rough-and-tumble play, and it is accompanied by a play face that seems cheerful to most humans. And since we are, after all, each other's closest relatives, sharing 99 percent of our genes and much behavior, there is a lot of intuitive crossover. Although chimp vocalization is a homologue of human laughter and will be referred to in this book as chimp laughter, it's important to distinguish it from its human counterpart. Such distinctions are essential for our comparative and evolutionary analyses.

The most notable acoustic similarity between human and chimpanzee laughter is its rhythmic structure. Whether a chimp is "pant-pant-panting," or a person is saying "ha-ha-ha," the sonic bursts occur at regular intervals, a property apparent in the waveforms of both vocalizations. Chimps, however, have a laugh rhythm about twice as fast as that of humans. (The chimp sounds were separated—onset to onset—by about 120 milliseconds, versus about 210 milliseconds for humans.) This is because the chimpanzees vocalize during both inhalation and exhalation. If only the more strongly voiced exhalation is considered, the chimpanzee laugh rate is halved and approximates that of humans.

Contrast the scruffy-looking, relatively structureless chimp spectrum with the sharply defined "ha-ha-ha" of human laughter. The chimp spectrum lacks the clear harmonic structure typical of the human laugh spectrum. (Recall that the harmonic structure of the human laugh refers to the regularly spaced stacks of frequency-bands, each of which is a multiple of a "fundamental frequency," the bottom

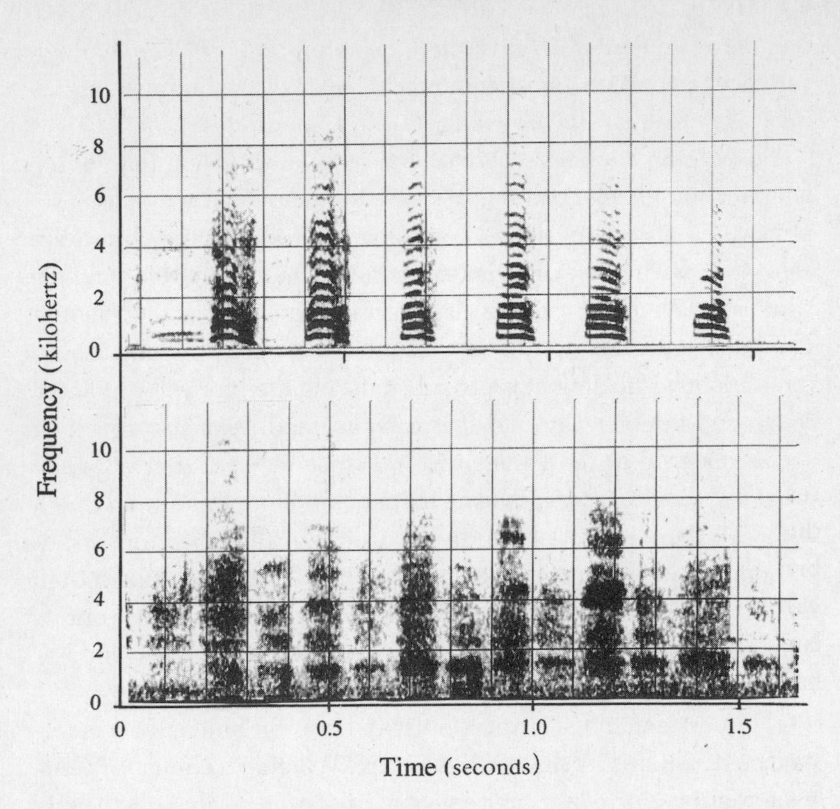

Figure 5.2 Frequency spectra of human (top) and chimpanzee (bottom) laughter. Breathy chimpanzee laughter lacks the distinct note form and strong harmonic structure (regular stacks of frequency-bands) characteristic of human laughter. (Adapted from Provine, 1996)

frequency-band of each stack.) Chimpanzee laughter also lacks the decrescendo of human laughter, the note-by-note trailing-off of loudness. Because chimpanzee laughter is panting, the volume of air available for a vocalization is renewed after each exhalation/inhalation cycle. This contrasts with purely expiratory human laughter in which the air for laugh-notes is gradually exhausted. Although chimpanzees never produce a humanlike laugh, humans can easily mimic the chimpanzee laugh.

Why Chimpanzees Can't Talk

Moving beyond acoustic analysis, we now contrast how chimpanzees and humans actually produce their own forms of laughter. A critical difference between chimpanzee and human laugh production is the relation in each species between vocalizing and breathing. The nature of this linkage explains why chimpanzees can't talk and people can. As described here, laughter, therefore, becomes a powerful neurobehavioral probe into vocal evolution

Human laughter, like speech, is produced exclusively during an outward breath—the discrete notes of laughter ("ha") are produced by chopping a single exhalation. Chimpanzee laughter, in contrast, resembles panting, with a single breathy vocalization being produced during each exhalation and inhalation. This coupling between breathing and vocalization was obvious from visual inspection and was confirmed by placing our hands on the heaving abdomens of the laughing chimps. If you want to try this demonstration and don't have a chimpanzee handy, you can, being a primate in good standing, simulate its laughter. Place your hand on your lower abdomen and pant—huff-and-puff—at a rate of four or five cycles (one exhalation and inhalation per cycle) per second. The exact rate isn't important. Note the prominent abdominal heaving associated with each pant.

Now, contrast the abdominal pulsations of chimplike panting with the smoother, tonic contraction produced by forcefully speaking the humanlike "ha-ha-ha." (Although true laughter is not a matter of speaking "ha-ha-ha," doing so works fine for this demonstration.) In this demonstration, you have experienced the very different patterns of neuromuscular activity responsible for chimp and human laughter. These differences have important implications for understanding the evolution of laughter and speech.

We humans speak as we laugh, by modulating an outward breath. If chimpanzees likewise speak as they laugh, by producing one sound per exhalation and inhalation, we have identified an important and previously unrecognized constraint on the evolution of

Figure 5.3 Contrast of human and chimpanzee laugh production. The sounds of human laughter, such as "ha," are produced by interrupting a single expiration (arrow). In contrast, chimpanzees produce only one laugh sound, a pant or gutteral "ah," for every expiration or inspiration (arrows). Humans laugh as they speak, by modulating an outward breath. The close coupling between breathing and vocalizing in chimpanzees may partially explain the failed attempts to teach these animals to speak English. (From Provine, 1996)

speech and language in chimpanzees and other great apes. Imagine the restrictions on your own speech if you were limited to one syllable per respiratory cycle. Chimps are captives of an inflexible neuromuscular system that is still closely tied to the essential and ancient labor of breathing. Indeed, the respiratory-vocal coupling of chimpanzees may be as limiting to the emergence of speech in the species as the structure of the tongue, larynx, and vocal tract and may be more resistant to evolutionary change. A shift in respiratory-vocal coupling would require reprogramming the neural output to muscles and the emergence of a time-sharing algorithm regulating breathing and talking, not the more subtle structural alteration of the instrument of vocalization.

As a "thought experiment," imagine a chimp receiving a transplant of a human vocal tract and respiratory apparatus in exchange for its own. Assume that the procedure was perfectly executed and that the transplant was innervated by the chimp nervous system. Would the chimp be able to use it, learning to "play" this new mechanism to produce human speech sounds like a musician learning a second instrument? Probably not, because the new apparatus would still be controlled by the old, inflexible chimp nervous system.[1]

The woeful vocal speech competence of apes is a point of agreement in the often contentious debate about primate language. Although several researchers have reared baby chimpanzees in conditions similar to those of human children, their simian wards acquired hardly any speech. In the most successful effort, Keith Hayes and his wife, Cathy, home raised the infant chimpanzee Viki. After six years of exhaustive vocal training, Viki could manage only the marginally perceptible words "mama," "papa," "cup," and "up." The inability of chimpanzees to produce English words was once the basis of an underestimate of their general linguistic and symbolic competence. The bottleneck in the evolution of ape speech probably lies more in the domain of sound production than cognition and symbolic capacity.

Breakthroughs in human/primate communication occurred when methods were devised to circumvent the inadequate vocal apparatus of the great apes. Allan and Beatrix Gardner and Roger Fouts

[1]For a fictionalized account of such a procedure, consult *Ancient of Days* by Michael Bishop, in which Adam, a surviving member of the presumably extinct hominid group *Homo habilis*, gets surgery to alter his vocal apparatus. Benefiting from his surgically acquired speech, Adam goes on to become an eloquent and skilled player in contemporary human affairs. He even gets the girl! All that was holding back this primate was his deficient vocal apparatus and some grooming tips. Unfortunately, this science fiction treatment does not explore the problematic issue of the neural control of vocalization raised above. Neither Adam, the fictional *Homo*, nor contemporary nonhuman primates are likely to benefit so greatly from vocal touch-up surgery.

used American Sign Language (ASL) to communicate with the chimp Washoe. Francine (Penny) Patterson followed their lead and trained the gorilla Koko in ASL. David Premack worked with chimp Sarah using magnetized plastic symbols that could be strung together on a board. Sue Savage-Rumbaugh and Duane Rumbaugh trained the chimp Lana and the bonobo (pigmy chimp) Kanzi to press symbols on a large computer display or point to them on a tablet. Under the tutelage of Savage-Rumbaugh and Rumbaugh, Kanzi learned to respond to hundreds of spoken English words. (The bonobo or pigmy chimpanzee, *Pan paniscus,* is distinguished by having more humanlike vocalizations and social characteristics than the common chimpanzee *Pan troglodytes.*) Although all of these student apes developed impressive vocabularies, critics argue that they did not achieve the holy grail of "true language" with its requisite grammar and sentence structure, learning only mindless tricks. We will not engage this heated debate about ape language, focusing instead on a weak link in the chimpanzee's mechanism of vocal expression. We now explore the evolutionary events that permitted the flowering of human speech and language, and broaden the circle of considered species beyond the great apes—a line of inquiry that was prompted by the neurobehavioral probe of laughter.

Bipedalism, Laughter, and Speech Evolution

"In the beginning was the word" (John 1:1). But it is truer to say that "in the beginning was the breath," because all else in vocal communication is fashioned from it. To speak, or to produce any other vocalization, is to periodically override or modify our most basic need, breathing. Eating, drinking, and having sex collapse into insignificance if you can't catch your breath. The ability to override so vital a function as breathing in the service of sound making was a revolutionary event in neurobehavioral evolution. A second revolutionary event was the achievement of the even greater respiratory control necessary to produce speech. Evidence of this second critical transition in respiratory and vocal control comes from contrasting the laughter of chimpanzees and humans.

Among the chimpanzees, especially the more vocally facile bonobo (pigmy chimpanzee), we witness animals on the brink, but lacking that uncertain something that changes everything, that small increment that enables speech to flow forth, where before there were only simple calls and cries. The critical transitional steps to speech are difficult to envision, and may involve incremental changes that have large, nonlinear effects. The first speech sounds may have been like our first small, uncertain steps in walking, a stumbling forward, barely breaking a fall with our next, just-in-time step. Quickly, these first tentative steps gave way to bold, certain strides, leaving no trace of their genesis. The blocking, hesitancy, and repetitions of stuttering may be a consequence of this chaotic heritage, an involuntary "locking up" on the threshold of fluency, the amplified, nonlinear effect of a glitch in coordinating the complex, unwieldy neuromuscular mechanism of speech. Indeed, laughter itself may be a "stuttered" vowel sound. Although we may never discover the critical physiological event or primal moment in the evolution of speech, comparisons of chimpanzee and human laughter provide a tantalizing glimpse of what might have been.

As considered above, chimpanzee laughter is locked into the cycle of breathing, with one pantlike laugh-sound being produced per exhalation and inhalation. Chimps are unable to chop an exhalation into the discrete "ha-ha-ha"s of human laughter. What elevates this distinction in laugh-form above footnote status is that it reveals a fundamental inability of chimpanzees to modulate an exhalation, a critical condition for the production of humanlike speech. Humans laugh as they speak, by the virtuosic modulation of sounds produced by an outward breath.

Scientists have paid little attention to the neuromuscular mechanism of breath control, the respiratory engine of speech production, even in humans. By neuromuscular mechanism, I refer to the fundamental process of how neurons and muscles produce the *move-*

ments that ultimately translate into vocalizations, not the anatomical features of the vocal tract, a topic that has received more attention. The reason for the neglect of respiratory mechanism is understandable—the motor control of vocalization is not a sexy research topic. How many philosophically and psychologically oriented linguists lust after the details of thoracic anatomy or the patterning of motoneuron impulse traffic to the muscles powering and controlling breathing and speech? Yet without an anchor in the natural world of sound making, the discipline of linguistics degenerates into a theoretically driven study of synthetic issues—a battle of logicians. Speech is indeed unique in some respects, as claimed by many linguists and speech scientists, but it's no less dependent than other behavior on underlying motor mechanisms. And the analysis of these mechanisms is the key to understanding the evolution of speech.

The laugh-probe indicates that the chimpanzeelike laugh/speech mechanism with its high degree of respiratory-vocal coupling is the ancestral form. The human variant with its looser respiratory-vocal coupling evolved sometime after we branched from chimpanzees about six million years ago. Evidence of the primacy of the chimp form comes from the identification of chimplike laughter in orangutans and gorillas, apes that split off from the chimpanzee/human line several million years before chimpanzees and humans diverged.

The laugh-probe reveals no neat solution to the age-old, multifaceted enigma of speech evolution, but it does suggest promising places to look for clues. Because so much hinges on respiratory control, this literature is likely to offer novel leads. Of particular interest are the relations among the evolution of bipedalism, breathing, and speech, three events that seem, on first hearing, to be only distantly related.

I will begin with a few well-established points about the evolution of human bipedalism and speech before considering the relation between them. Humans evolved the ability to walk and run skillfully and efficiently in an upright position some time after we split off from an ancestor shared with chimpanzees. Chimpanzees and other great apes, although not obligate quadrupeds, cannot walk efficiently or for long distances in an upright posture. The evolution of bipedality was a critical event in our species' biological and behavioral history that had consequences ranging from the freeing of hands for carrying and gesturing, to the natural selection for more stride-efficient narrower hips (less side-to-side rotation), a transition that had substantial costs to females, who had to endure more difficult birthing. The evolution of speech was another critical event in human history that occurred sometime after our branch point with chimpanzees.

I propose that *the evolution of speech and bipedal locomotion are causally related,* the basis of what I call the *Bipedal Theory* of speech evolution. (It was the "Walkie-Talkie" theory in one media account.) The common link between both acts is *breath control.* The evolution of bipedalism set the stage for the emergence of speech by freeing the thorax of the mechanical demands of quadrupedal locomotion and loosening the coupling between breathing and vocalizing.

"Running and Breathing in Mammals," a *Science* magazine article by Dennis Bramble and David Currier, spurred my interest in this topic. The authors provided comparative information about running and breathing in a variety of animals, including human joggers. They report, for example, that quadrupedal species such as horses usually synchronize their locomotor and respiratory cycles at a constant ratio of 1:1 (strides per breath). This synchronization is understandable because both locomotion and respiration use cyclic movements of the thoracic complex (sternum, ribs, and associated musculature). In addition, during quadrupedal running, the thorax is subject to powerful, repetitive impacts as the forelimbs strike the ground. Some sort of anatomical support or respiratory maneuver is necessary to strengthen the otherwise structurally weak, air-filled bag of the thorax. Humans increase the rigidity of their thoraxes by breath holding when rising from a chair without using their hands, or when lifting a heavy weight. And they may "bear down" and grunt if a task is really challenging. This act of breath holding and bearing down, which is called the Valsalva maneuver, is also used during defecation, and if you pinch your nose shut, is handy in inflating your middle ear to relieve pressure when scuba diving or landing in an aircraft.

Human runners are unique among modern mammals in having a striding bipedal gait. Our upright posture and locomotor pattern frees us of forelimb impacts and loosens, but does not abolish, the coupling between running and breathing. In contrast to the usual 1:1 ratio of quadrupeds, bipedal human runners employ a variety of phase-locked patterns (4:1, 3:1, 2:1, 1:1, 5:2, and 3:2), with 2:1 as the most common pattern. Here we have evidence of the plasticity of the human respiratory rhythm and its relation to upright bipedal gait. At this point, Bramble and Carrier's insight triggered one of my own—that bipedalism was necessary for the evolution of speech.

Bipedalism was a necessary, although probably insufficient condition for the evolution of speech in primates. Further embellishments of the vocal system were necessary, but bipedalism was a critical first step.

Further support for the bipedal theory came from a chance meeting with Ann MacLarnon, a primate spinal cord expert, at the 1996 Congress of the International Primatological Society in Madison, Wisconsin. Ann commented that my detection of respiratory limits of chimp vocalization may be related to spinal cord structure, adding that the main difference between the spinal cords of humans and other primates is that humans had more spinal cord mass in the thoracic segments controlling the neck, arms, and trunk. This is the cord region implicated above in the evolution of human respiratory control. In addition, the greater size in humans exclusively involved gray matter, the area of the spinal cord composed of neuronal cell bodies, circuits that pattern movement, and motoneurons that send nerves to the muscles. No difference was observed in white matter, the cord area composed of nerve fibers that transmit information within the nervous system.

After the meeting, I found another of Ann's contributions: her analysis of fossil remains of the Nariokotome *Homo erectus*, an extinct bipedal hominid discovered in northern Kenya. Based on inferences about the size of the hole in the fossil vertebra that sheath the spinal cord, she determined that *erectus* had a small-diameter thoracic spinal cord characteristic of nonhuman primates. Thus, the more massive human thoracic spinal cord is not a response to the challenges of upright posture or bipedal gait, an event that occurred several million years before the appearance of *erectus*. The bloom in human thoracic spinal mass was the consequence of other demands, perhaps the respiratory control of the muscles necessary for the evolution of speech. Taken together, the comparative study of primate spinal cords and the fossil record of *Homo erectus* suggest that bipedalism appeared before speech, a sequence consistent with the evolutionary scenario advanced in this chapter.

Does bipedalism signal the presence of the respiratory/locomotor/vocalization coupling necessary for speech or exceptional vocal competence? Human developmental studies offer provocative, indirect support for such a link between locomotion and vocalization. Standing alone and walking is a significant developmental milestone reached around the end of the first year of postnatal life, a birthday celebrating the acceleration of life's next great task, the development of speech and language. (The shaping of vocalization, however, starts much earlier.) Although the development of bipedal locomotion is obviously not necessary for the onset of speech (speech develops in children who never walk), it may signal the typical maturational stage when the neuromuscular mechanism for speech is in place. Children need not walk to benefit from their species' heritage of bipedality.

Does the proposed bipedalism/respiration/vocalization link hold for animals other than primates? Unfortunately, lessons to be learned from comparative analyses are limited because no other mammals either speak or routinely walk on two legs. And distantly related animals may evolve means other than bipedality to loosen respiratory constraints on vocalization.

The birds, virtuoso vocalizers and sometimes imitators of human

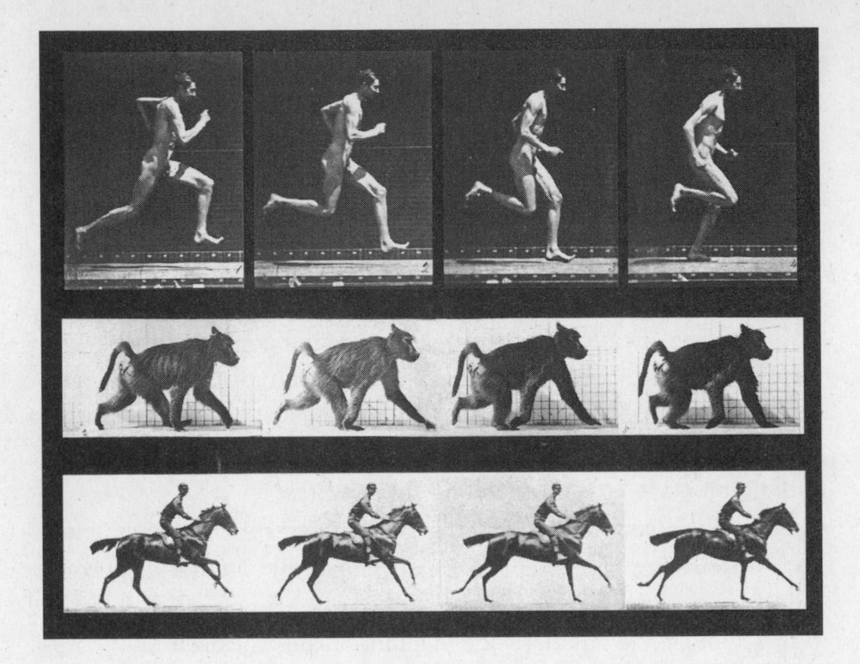

Figure 5.4 The evolution of bipedal locomotion. The emergence of bipedalism in humans' ancestors facilitated the evolution of speech by freeing the thorax of the mechanical demands of quadrupedal locomotion. Quadrupeds must synchronize their locomotor and breathing cycles to increase the rigidity of their air-filled thorax to absorb the powerful forelimb impacts during running. In contrast to the 1:1 ratio between breathing and striding in quadrupeds, bipedal human runners employ a variety of patterns, evidence of looser coupling between running and breathing. The emancipation of breathing and vocalization from locomotion was necessary for the vocal virtuosity of human speech. (Adapted from Muybridge, 1899/1957, 1901/1955)

speech, are fellow bipeds that are obviously able to redirect breathing for sound making. But birds are so specialized that they may be a unique case—the avian pectoral apparatus and forelimbs have been remarkably transformed in the service of flight, and their vocal apparatus is radically different from our own. And avian vocal competence is hardly uniform—for every mynah or songbird, there is a duck or fowl. It's significant that the African gray parrot, a bird noted for its expert mimicry of human speech and other sounds, also has remarkable symbolic competence. In some birds, at least, vocal facility may be a step toward the emergence of language and other higher-order cognitive processing.

Whales are a fascinating class of accomplished mammalian vocalizers that circumvented the bipedal phase of vocal evolution. The respiratory apparatus of these gigantic descendants of terrestrial quadrupeds is not freed of the demands of gravity by bipedality, but by suspension in water. And whales are also capable of remarkable feats of breath holding for tens of minutes, obvious evidence of breath control, another possible behavioral marker of vocal potential. We humans, of course, can hold our breath for the less impressive, but still significant interval of one to several minutes. But breath holding by itself does not indicate vocal facility—ducks and other diving birds are not known for their song.

So what about the vocal competence of other breath-holding mammals? Hoover the seal is a remarkable case. This human-raised male harbor seal is reported to speak about a dozen English words, and even produce a rousing belly laugh. (The "words" may have been combinations of assorted vowellike sounds characteristic of seals.) Although I've not heard Hoover's laugh and could not detect discrete words on two audiotape recordings of his vocalizing, I was impressed by the speechlike quality of what I heard. (Imagine the gruff voice of an inebriated and incoherent male streetperson accosting you on the sidewalk.) Sadly, these feats of seal sound making can't be pursued because Hoover died several years ago. Perhaps another seal of vocal distinction will emerge and carry on Hoover's honorable tradition.

The present approach to the origin of speech is much nearer its beginning than its end and has gaping voids. Although the evidence relating laugh pattern, respiratory coupling, and speech competence in humans and chimpanzees is well established, other topics are less well developed. What, for example, are the specific neurobehavioral changes associated with bipedalism, and how, exactly, do they nudge the vocal apparatus toward speech production? But the bipedal hypothesis and associated data answer significant questions and may stimulate creative sleuthing for clues at the intersection of breathing, walking, talking, and laughing.

Ape Laughter and Paleohumorology

After wrestling with the origins of laughter and speech, this chapter concludes with deliberations about an issue of the here-and-now, the matter of chimp humor. What, exactly, makes chimps laugh, and does this knowledge suggest the nature of the most ancient joke? The problem is deceptively difficult. Speculation about the nonhuman psyche is one of the most treacherous and notorious pursuits in the history of psychology. Although chimpanzees are the animals most like ourselves, they differ fundamentally from us in their lack of spoken language, and lacking language, they probably lack the capacity for abstract thought necessary for most humor as we know it. To minimize the anthropomorphic biases inherent in the search for humor, our tactic will be conservative and austere, avoiding inferences about the machinations of the simian psyche. We will focus instead on the stimuli and behaviors that immediately precede laughter, the approach used previously with humans. After considering the stimulus context of infant and adult chimp laughter, the chapter will end on a more speculative note with anecdotes about ape behavior that may meet human standards for humor.

Solitary chimps, like solitary people, seldom laugh, a result consistent with laughter's role as a social signal (Chapter 3). But chimps and humans differ in the social situations in which they laugh—we humans have added something new to our still present chimplike tendencies. Adult humans laugh most during conversation. Chimps, in contrast, laugh most when tickled, during rough-and-tumble play, and during chasing games (the chimp being chased laughs most). Physical contact or threat of such contact is a common denominator of chimp laughter. Nevertheless, chimp laughter is probably less social than in humans—I have observed solitary young chimps laughing while playing with objects (balls), tickling their own feet (Can they tickle themselves?), rolling on the ground, and swinging. Roger

Fouts concurs that chimps sometimes laugh during solitary play, recalling a particular case when Moja was playing with old clothes. Chimps lack a clear equivalent to human conversational laughter in which two physically separated individuals look at each other, gesture, or vocalize before breaking up in a fit of laughter. The physicality and social context of chimp laughter resemble that of human children before the age of five or six when joking becomes prominent and intentional. Although chimps and young children engage in acts that produce laughter in others, such laughter is typically the incidental by-product of play, not its objective.

In a rare field study of the interactions between chimpanzee mothers and infants, Frans Plooij made an important discovery about the roots of laughter. Baby chimps control the behavior of their mothers, with tickle and laughter playing significant roles in a nurturant pas de deux. Compared to their human counterparts, chimp mothers rarely gaze at their babies or spontaneously engage in communication, but they do respond to their babies' actions. The chimp baby initiates mother-infant play by biting the mother, who then looks at and tickles the baby, triggering cycles of biting/tickling interplay that continue until the baby signals "too much" by "defending," fussing or crying. The baby signals the "just about right" amount of stimulation with a "play face" and laughter. In these duets, the *baby*, not the mother, initiates and regulates the interaction. Although human mothers are more proactive and attentive, laughter, smiling, and tickling retain a central role in the social interplay, with babies regulating the intensity and duration of the interaction. Laughter and smiling are means for a prespeech human baby to indicate, "I liked that, do it again," and fending away and crying are signals for "Enough!" or "Too much!"

Chimpanzee mothers rank below humans and above monkeys in attentiveness to their babies. Monkey mothers, for example, don't respond to their babies' bites like chimp mothers, but sibling and peer monkeys do. The responsiveness of these young monkey companions to biting babies may explain their effectiveness as peer "therapists" in Harry Harlow's developmental studies of baby rhesus monkeys deprived of their mothers. Whether due to ineffective play face and laughter, or to an unresponsive parent, monkey babies lack the stimulus tools necessary to capture and hold their mothers' attention in the social give-and-take typical of chimpanzee and human mothers.

Most prelaughter stimuli for chimpanzee laughter are physical, concrete, and "nonjoking." Consider stimuli for chimp laughter that may meet human standards for humor. (This criterion may not be terribly high—after all, it includes the Three Stooges, Howard Stern, and *Married with Children.*) The enterprise is hampered at the outset by the failure of chimps to signal their production or perception of humor with laughter. Although much human humor does not trigger laughter of a human audience, enough of it does to establish a pattern that can be generalized to similar, more subtle titillations that don't reach the threshold for guffaws. We are not so fortunate with the otherwise playful and laughing chimps. Without the "gold standard" of laughter to gauge the occurrence of chimpanzee humor, we must make inferences about the mental life and intentions of another species, an ill-advised pursuit. There are no systematic studies of ape humor to guide our way. The evidence is found in scattered, anecdotal reports by researchers and caregivers based on homocentric estimates of whether a given primate act meets the human criterion for "humor." Having been forewarned, let's examine some of this provocative but necessarily limited data.

Most candidates for simian humor involve cases of intentional misusing of objects and misnaming of people and things. For example, researcher Roger Fouts observed the signing chimpanzee Washoe using a toothbrush as if it was a hairbrush. Moja, another of Fouts's signing chimpanzees, called a purse a "shoe," put the purse on her foot and wore it as a shoe. Francine "Penny" Patterson observed the signing gorilla Koko treating rocks and other inedible substances as if they were foods, offering them as "food" to people.

When Koko was asked by Penny to feed a baby doll with a bottle, Koko held the bottle to the baby's eye instead of its mouth. In a misnaming incident, Koko refused to give the sign for drink (extended thumb with hand in fist position as if you are drinking from a thumb), finally providing the proper sign except that the thumb was placed in the ear instead of the mouth. In another misnaming incident, Koko spontaneously signed "bird" to a picture of a bird. But when asked later what the picture was, she signed "flower." The above cases of presumed intentional "misnaming" and "misusing" are potential jokes typical of human children of preschool age. Reports that the apes appeared to be in a playful mood, or glanced at the caregiver for evidence of the effect of their errant actions, suggests but does not establish a joking intent. Another widely noted class of misnaming involves "name calling," an act probably more amusing to human witnesses than the provoked primate signers. When upset with her caregiver, gorilla Koko referred to her as "dirty toilet." Lucy, a chimp reared by Maury and Jane Temerlin, signed "dirty cat" after a conflict with a local feline, and "dirty leash" in regard to a disliked restraint. But are such misusing/misnaming cases errors, lies, jokes, or simply misinterpretations by caregivers?

In another possible instance of simian humor, Roger Fouts reported that while riding on his shoulders, the chimpanzee Washoe urinated on him, signing "funny" (touching her nose) and snorting, but not laughing. This incident is notable both for Washoe's reported recognition of a potentially funny event and the absence of laughter following the humorous episode. Fouts provides a second anecdote, this time suggesting an association between laughter and the act producing it. Instead of signing "You tickle Booee," the chimp Booee signed "You ——— Booee," substituting the sound of laughter ("pant-pant") for the command to tickle. In different ways, do Washoe and Booee show a primitive appreciation of the concept of "funny," the name we humans give to stimuli that make us laugh?

The sparse evidence of primate humor suggests a cognitive asymmetry whereby apes may appreciate their own self-initiated humor such as misnaming, but are bewildered by similar misnaming by others (e.g., chimp Lana), or respond to it by signing "stupid" (e.g., gorilla Koko), a response like that of human preschool children who often complain that jokes they don't "get" are "stupid" or "silly." The apes and preschoolers may lack the theory of mind necessary to infer the playful intent of others. Thus, humor may evolve in the perpetrator before the audience or victim. Here we find support for the common view of the primitiveness of the practical joke and joker.

Whatever the style of humor, alcohol primes the laugh mechanism of chimpanzees as it does in humans. Consider the report of Maurice Temerlin, who likes to share a cocktail or two with his home-reared chimpanzee "daughter" Lucy. "In some ways, Lucy is an ideal drinking companion. She is very appreciative, always making sounds of great delight when offered a drink. She never gets obnoxious, even when smashed to the brink of unconsciousness. Alcohol relaxes her and it improves her sense of humor, for she laughs, tickling herself, posturing before a mirror, making 'crazy' faces and laughing at them."

Comparative humorology is an endeavor fraught with even more problems than the controversial study of ape language. But after sharing the anecdotes of others, I am emboldened to offer my own best guess about the most ancient "joke" and the origin of the laughter it triggered. My entry into the paleohumor sweepstakes is conservative, nonlinguistic and cognitively impoverished—*feigned tickle*. Although this simple physical act may fall short of common conceptions of humor, it clearly involves "kidding," playful intent, and triggers laughter. The power of "I'm going to get you!" and its variants to evoke laughter is undeniable—young chimpanzees and humans remain suckers for this ancient and invited ruse. Unfeigned, normal tickle is an even more primal stimulus of laughter and a likely form of "protohumor," but I didn't nominate it because the response of the ticklee is more reflexive. And what was the origin of the primal laughter in these ancient ticklefests? As suggested earlier, the vocalization of laughter did not arise de novo, but originated in the ritualized panting of rough-and-tumble and sex play, whereby the sound of labored breathing came to symbolize the playful state that produced it. The vowellike "ha-ha-ha"s that parse the outward breath in modern human laughter is one step removed from the archetypal huffing and puffing that signaled laughter and play in our ancient ancestors.

SIX

Ticklish Relationships

If you tickle us, do we not laugh?

—Shakespeare, *The Merchant of Venice*

olo tickle is even emptier than solo sex. Most adults can masturbate to climax, whereas self-tickle is a pale shadow of its social counterpart. You can no more tickle yourself to laughter than you can startle yourself. Tickle (synonyms: kittle, vellication) is the product of a social interaction between a tickler, the person or thing administering the stimulus, and the ticklee, the person being stimulated. Just as social interactions are the key to understanding tickling, tickling is essential to understanding the associated social vocalization of laughter. As we will see, tickling is not a footnote in the story of laughter but at the very heart of it. If tickle seems a divergence, consider that compared to comedy, tickle is a more reliable and much more ancient stimulus for laughs. Tickle and associated rowdy play are the main sources of laughter in our primate relatives and are the ancestral stimuli for laughter (Chapter 5).

Tickle reveals why we laugh "ha-ha" instead of making some arbitrary sound such as "ba-ba" or "tsk-tsk." New evidence provides startling insights about who tickles whom and why, and forces a rethinking of the origins of laughter, play, and the biological basis of social behavior. Along the way, we consider social touch, sex play, a quirky sexual fetish, a mechanism for autism, and the neurological basis for "self."

For millions of years people have tickled and been tickled, but scientists have paid little attention to all of this tactile friskiness. As a research topic, tickle is even lower in the scientific pecking order than laughter and it has a much smaller scientific literature. And this literature is not always enlightening. The distinguished multivolume *Handbook of Physiology* relegates tickle to the chapter on pain, where author William H. Sweet notes that the distinction between tickle and itch "scarcely seems worth fussing about." Tickle will not yield its secrets to anatomists, physiologists and others who neglect the critical social dimensions of the stimulus.

Tickle involves more than the sensory physiology of touch and the physical properties of the stimulus. Contrast, for example, the sensory experience of having your ribs caressed by a lover versus a stranger. The identity of the bearer of a physically identical feathery touch makes the difference between a ticklish delight and an ordeal, an insight recorded by Darwin over a century ago ("A young child, if tickled by a strange man, would scream from fear."). Social relationships are the key to understanding the enigma of tickle. An individual's inability to stimulate himself to laugh and experience the sensation of tickle is one of the few points of agreement about this curious behavior, and is the essential starting point of our analysis. Tickle is not a simple, brief, reflex of the knee-jerk variety that can be evoked equally well by you or a stranger encountered on the sidewalk. The response to tickle is an innate, socially and behaviorally complex reaction directed toward terminating the stimulus that triggered it.

The more I talked with friends and colleagues about tickle, the clearer it became that this was no trivial behavior. Some people liked to be tickled, and others hated it, but hardly anyone was without an opinion. A few women became upset just talking about tickle and considered tickle always to be a highly offensive intrusion into their personal space. The *Child Safe Program* manual sent home with Baltimore area parochial school children warns against "tickling too long," a kind of "confusing touch" that could trigger charges of child sexual abuse against the errant tickler. In her newspaper advice column, Abigail Van Buren ("Dear Abby") expresses deep concerns about a father who tickles his young daughter "too much." Tickling girls is even illegal in Norton, Virginia. Tickle is becoming a casualty of our touch-averse and abuse-sensitive society.

The tickle project began with a questionnaire to tap people's lifelong experiences with tickle. In such a study it's important to cast the broadest possible net of inquiry to insure the detection of key variables and the pursuit of appropriate questions. It's presumptuous to assume beforehand what is important to know about a topic, particularly one as understudied as tickle. This adventure in tickle science was conducted in collaboration with Bernie Fischer, who is now a student at the Medical College of Virginia.

In our 52-item questionnaire, people were asked who tickles whom and what they think about tickling and being tickled. The 421 respondents (146 males, 275 females) ranged from 8 to 86 years of age and included students, neighbors, relatives, friends and their children, friends of friends, and a few random acquaintances.

Tickle passes the critical test of *natural validity* (Chapter 1)—its frequent occurrence suggests that it has a significant role in our lives, an important criterion for scientific merit. Thirty-five percent of our respondents *had been tickled* during the past week, 86 percent during the past year. These respondents were not the victims of roving bands of crazed ticklers: they know who tickled them, and 40 percent of subjects themselves *had tickled* someone during the past week, 84 percent during the past year.

Being tickled was moderately pleasurable to most respondents, and many ticklees liked it a lot, a startling conclusion to confirmed tickle haters. On average, being tickled was rated 5.0 on a scale ranging from 1 ("very unpleasant") to 10 ("very pleasant"). (The middle value on a 10-point scale is 5.5.) As we will see later, this moderate pleasure rating is consistent with the most common motive for tickling and being tickled, "to show affection." And tickle bouts are filled with laughter, the signal of many pleasant encounters. Most ticklees (84 percent) and almost as many ticklers (80 percent) reported that they laughed during tickle episodes. Such laughter is important in maintaining the reciprocity of the tickle relationship. "I dislike tickling someone when it doesn't make them smile or laugh," noted a 20-year-old female.

Being tickled was slightly less pleasant to females (average rating of 4.8) than males (5.4), with about twice as many females as males (16 percent versus 9 percent) rating the experience as "very unpleasant." Serious gender conflicts can arise from such differences. A well-meaning and insistent male tickler could earn a knee in the groin delivered by a revulsed female ticklee, leaving the clueless perpetrator wondering what happened to the magic in their relationship. Sometimes communications are clouded, as with the 21-year-old female who reported that "My brother (three years older) used to 'torture' me and I would scream and yell and pretend to be mad, but actually I loved the attention and the laughter."

Although ticklees may enjoy the experience, they typically engage in a defensive reaction that may include withdrawing the tickled body part, fending away the hand of the tickler, and hunching over and pulling in the crooked arms to protect the ribs ("huddling"). Less often, children will solicit tickle by guiding the hand of the tickler to the site to be tickled. The common fending-away of the tickling hand is not a uniquely human gesture—I have observed it in chimpanzees and gorillas. During tickle, we find other hints of coyness and approach/withdrawal, evidence of tickle's heritage as a defensive behavior. Children, the most enthusiastic tickle lovers, often run away shrieking from the tickler, only to return seconds later for another dose of tickle and laughter—this is their way of regulating the duration and intensity of tickle. Babies can't run away; they resort instead to laughter, fussing, or crying to titrate the intensity of tickle.

The most aversive tickle bouts in children and adults are those in which the ticklee is held down, helpless to regulate the onslaught. The ordeal of uncontrollable tickle stimulation is rooted in the function of the tickle response—an all-out, emergency reaction to get the stimulus to stop, a reaction that serves the ancient and vitally important role of defending the body surface. The urgency and occasional violence of ticklee behavior is much more akin to the struggle to withdraw one's hand from a fire than scratching a nagging itch. We can sympathize with the 20-year-old female who reported, "My dad used to . . . hold me down while he and my younger brothers tickled me till I almost cried. I screamed and yelled and hated it!" Forced tickle was enjoyed by almost no one—an exception to the otherwise varied tastes of ticklees. But there were rare exceptions, as with this 18-year-old female. "As a child, I remember me and my dad ganging up on my brother and then [them] ganging up on me. We would have tickle wars and it makes me smile and laugh to remember the fun we had!"

Tickling has powerful physiological effects, some of which are unpleasant. "My brother used to tickle me so much that I would have asthma attacks, and that is why I hate to be tickled," noted a 21-year-old female. Other body systems can be affected. A 10-year-old boy reported, "I was tickled by my dad and it made me pee in my pants"—a most humiliating experience, and one shared by several other respondents. But not all extreme states of tickle-induced physiological arousal are negative. "My boyfriend and I tickle each other all the time until we are out of breath and coughing from laughter" (23-year-old female). And "When I was young, my father would always tickle me and make me laugh until I could hardly breathe. I *loved* it!" (22-year-old male).

Many tickle haters base their negative evaluation of tickle solely on extreme and obviously unpleasant experiences, acts that share features with rape and other nonconsensual and nonreciprocal sexual contact. A 20-year-old female observed, "If the tickler is a person I don't know well, I feel violated, like the person is molesting me." But just as rape is not typical sexual behavior, neither is forced stimulation a representative tickle experience.

The sometimes aggressive role of the tickler is reflected in our vocabulary. Although tickle is usually a nonadversarial, consensual encounter with family and friends, the ticklee is often referred to as the "victim" of a tickle "attack." Despite this language of conflict, many people speak of cherished childhood memories of family "tickle battles" and continue the tradition of tickle combat as adults with their own partner and children. "My sister and I would lie in wait beneath the covers for my parents to go to bed. When they got into bed, we would launch a 'tickle attack.' Everyone would roll around, tickling each other and laughing, having a wonderful time" (21-year-old female).

Children also enjoy another pseudo-aggressive tickle variant, the "tickle monster" game in which an adult or another child pursues victims who run laughing and shrieking in mock terror. Chimpanzees enjoy similar tickle/chasing games, with the chimp being chased doing most of the laughing (Chapter 5). Russian mythology is populated with "real" tickle monsters to fire children's fears and imagination. The *leshii,* malevolent master spirits of the forest, sometimes lure victims away from the path and tickle them to death. Fatal tickle is also dealt by the *rusalki,* water dwelling spirits of drowned maidens who snare victims with their naked beauty, laughter, and song.

In tickle bouts, it's more blessed to give than to receive. Tickling earned an average pleasure rating of 5.9, versus the 5.0 rating for being tickled on the 10-point scale. Even those who dislike being tickled often show at least modest enthusiasm for tickling someone else. As some women may suspect, males enjoy tickling a bit more than females (6.1 versus 5.9). But the moderately high pleasure rating of all parties (ticklers and ticklees of both sexes) in the tickle relationship indicates why tickle persists—most people enjoy it.

If you doubt the enthusiasm for tickling, consider the Tickle Me Elmo doll, the toy superhit of the 1996 Christmas shopping season. During December 1996, the news was filled with accounts of stampedes and fights to buy this fat, fuzzy, red doll. If you press Elmo's belly, he laughs, then says, "That tickles." On second press, he laughs, followed with, "Oh, boy!" On third press, Elmo simultaneously laughs and vibrates (simulating a struggling ticklee), concluding with, "Oh, boy! That tickles!"

In Elmo, we have an intriguing artifact of pop culture that taps several of our biological predispositions. Like most dolls, Elmo is "cute," exhibiting those cues of immaturity that trigger nurturing behavior of many species toward their young. But of special interest here is Elmo's laughter and struggling response to human "tickling," attractive stimuli to many humans, including a 19-year-old female student who noted, "I like to tickle little babies on their feet because they laugh and squirm." Elmo's laughter may even trigger pleasure and a contagious laugh response in the tickler. One 22-year-old female observed, "I love to tickle my boyfriend because he has a great laugh, and it makes us laugh together." Whether by chance or design, there is a lot of science in Elmo.

The strangest variation on tickling was discovered by one of my graduate students while pursuing wisdom and truth on the Internet. She e-mailed me a copy of a request by a woman (gender on the Internet is conjecture) who liked to watch people being tickled and would back up her desire with cash. The advertiser sought 30- to 60-minute videos of "very intense, shirtless, sockless, tied-up tickle torture" of young men (under 25) being tickled by one or more women. Enterprising videographers could collect $125 to $250 for their effort. I wondered what use was being made of this very special video collection. This discovery triggered a Web search that was the start of an intriguing but kinky adventure into the world of sexual fetishes and sadomasochism.

If you are not offended by sometimes X-rated tickle esoterica (and erotica), and are over 18 or 21 years of age (the standard disclaimer), you may check out this material by doing a Web search for "tickle" or "tickling." Tickle haters who view all tickling as a form of domination will find a lot of support here. What you will find ranges from graphics to favorite tickle stories, mostly involving women tickling women, with the ticklee experiencing varying degrees of undress and restraint (bondage). One particularly rich but now defunct Web site provided subsections devoted to favored tickle regions (feet, knees, tummy, ribs, armpit, upper-body, etc). There are aficionados of toe-sucking, armpit nuzzling, and everything in between. Pursuit of region-specific tickle led me to the surprisingly extensive Web coverage of foot fetishes (e.g., *In the Feet of the Night, Solefully Yours*), a good-natured, if nontraditional turn-on overlapping with tickle fetish sites.

A striking feature of stories provided by these tickle fetish Web sites is that tickle and associated, often unconsummated, sexual arousal, is the central theme. Early stages of these stories are typical of the standard porno flick. But unlike the real-life experiences of some of my questionnaire respondents, these often nude tickle frolics didn't necessarily lead to frenzied copulation, just more tickling and laughter.

The tickle sites' stories and graphics generally lack the give-and-take ("tickle battles") of normal rough-and-tumble and sex play, not surprising given the limited options of the often shackled ticklee. Most tickle scenarios take the tickler's perspective, an orientation consistent with my questionnaire's finding that it's usually more fun to tickle than to be tickled. (In the language of the sadomasochists, there is more interest in being a "top" than a "bottom.") Presumably, most site visitors are male—the ticklees shown or described are usually female, and my research discovered a very strong heterosexual bias in tickle relationships. The mostly female ticklers are probably a necessity of staging. Someone has to do the tickling, and heterosexual men don't want uninvited men showing up in their sexual fantasies. The additional women are less threatening, and perhaps further arousing, to the testosterone-addled male psyche.

For many enthusiasts, tickle is a form of "soft S and M," with discipline wielded by fingertip and feather, not whips and chains—dominance with training wheels. The tie-in of the kinky fun of tickle with the shadowland of S and M, dominance, bondage, and the like is not fanciful—Web sites devoted to these more disturbing themes are often cross-linked with those for tickle. With the exception of manic laughter, the struggling ticklee resembles victims of less benign torments.

"A rare case of sadomasochism (tickle torture)" is the report in a clinical research journal of an informative case of sexual fetish that is probably not as unusual as the title suggests—torture is less likely to be reported if administered by feather than by cat-o'-nine-tails. The patient is a 39-year-old lawyer who from childhood was sexually obsessed with tickling, especially watching, reading, or thinking about nude victims being forcibly tickled. His life was dominated by his sexual fantasy, and he fervently collected tickle-related art and literature of the sort displayed by fellow connoisseurs at the Web sites. Although his first sexual interests were primarily homosexual, he eventually married and had a child with an understanding and resourceful woman who cooperated with his obsessions. When wife tickling lost his sexual interest, his wife invited friends to her house and tickled them while her husband watched surreptitiously from another room; she then abandoned her guest to join her now aroused husband in intercourse. A variant of this scenario involved a servant girl rushing in and jokingly tying up the guest, removing her shoes and stockings to bare the soles of her feet, while husband and wife watched and had sex in another room. It's not surprising that "in the course of time, all of their friends stayed away."

Science is particularly rewarding when the many pieces of the empirical puzzle fit together, providing the pleasing aesthetic of closure, and confidence that tickle is being studied in an appropriate manner. The relation between tickler and ticklee provides some particularly useful pieces of the tickle puzzle. Unifying themes emerge when we add faces, names, and motives to players in the tickle game.

During the past year, who tickled *you* the most? Overwhelmingly, my respondents were tickled by family and friends, the same group of people upon whom they conferred varying degrees of *skinship,* the privilege of intimate touch. Hardly anyone (0.2 percent) was tickled by a stranger—the specter of being assaulted by a nameless person bearing a feather is not the stuff of everyday tickle experience. The immediate and extended family (e.g., husband/wife, mother/father, brother/sister, son/daughter, uncle/aunt, nephew/niece, grandfather/grandmother) tickled 42 percent of respondents at least once during the previous year. A 19-year-old male college student captured the familial nature of tickle and the maturation of its expression. "While a young child, it was a bonding experience with my family. Now it is a loving way of play with my significant other."

Ticklish relationships have a strong heterosexual character. When asked who tickled them most during the previous year, my mostly adolescent and adult respondents were more than seven times as likely to name someone of the opposite than same sex. If a friend of the opposite sex tickles you, there may be romance in your future, or at least some petting. "Tickling is the perfect way to touch girls with a good excuse," suggests a 20-year-old male. "If you accidentally touch their private parts, you can always say, 'Oops, I'm sorry, I was just tickling you.'" This tactic was not lost on the 22-year-old female who noted, "The person may want to touch the other person romantically, but doesn't know how or when to start, so he/she tickles the other person to initiate the touching." By far, the highest incidence of tickling involved those with whom the respondents had a romantic interest or relationship. "At a swimming pool, there was a boy who was interested in me. Instead of talking to me, he tickled me" (22-year-old female). "On my first date with my girlfriend we were fooling around and I tickled her. She retaliated and that was the start of a wonderful relationship" (21-year-old male).

When asked "Who would you most like to tickle you?" the heterosexual bias really escalates. In this idealized world, extraneous ticklers, those pesky aunts and uncles, brothers and sisters, etc., disappear, and we are left with the distilled essence of tickle lust. The over seven-fold heterosexual majority in the real world of tickle jumps to 15-fold when people are asked whose tickling they would most like to experience. These are huge effects in the chaotic world of behavioral science.

The question of tickler preference is useful in breaking free of rigid preconceptions about tickle. Even confirmed tickle haters can conceive of situations in which tickle may not be half bad—and perhaps very good, indeed. When asked when she most liked to be tickled, a reluctant 23-year-old female noted, "Never, really, but with a boyfriend in bed is OK." Her college classmate was more enthusiastic: "When my boyfriend tickles me, anytime, anywhere" (21-year-old female). If the link between tickle and sex seems wildly speculative, consider that the Dutch word for clitoris is *kittelaar,* "the organ of being tickled or titillated," and there is also the quirky evidence of the tickle fetish Web sites.

When the above questions from the perspective of the ticklee are flipped around and phrased from the perspective of the tickler, the results looked roughly like those just described for ticklees. When asked, "Who did you tickle during the past year?" and "Who would you most like to tickle?" respondents showed a heterosexual bias of over three- and five-fold, respectively.

In summary, the 421 respondents strongly agree that they tickle and are tickled by relatives, friends, and lovers, people with whom they have close social bonds. But does tickle play a role in establishing and maintaining these relationships? Or do people simply tickle those whom they are around the most? To answer these questions, the respondents were asked to select all applicable statements from a list of 10 options about tickling and being tickled.

By far the most common reason for being tickled was "to show affection" (72 percent). (The total of all responses exceeds 100 percent because respondents were instructed to choose "all that apply.") Fewer thought "to annoy" (45 percent) or "to get attention" (41 percent) was the motive of ticklers. A substantial percentage (27 percent) of respondents thought that "being tickled sometimes hurts or is painful," and others noted that being tickled "makes me angry" (14 percent), "is an invasion of my privacy" (9 percent), or "is a form of assault" (6 percent). Nevertheless, only 7 percent thought that "it is never appropriate for someone to tickle me."

Results from a complementary set of opinions asked from the perspective of the tickler were generally comparable to those of the ticklee, although there was a lower percentage of negative attributions ("to annoy," etc.) when the respondent was the tickler. We are generous when rationalizing our own motives. Some respondents were downright philanthropic in their tickling. "I like to tickle when I hold an upset child to cheer him up" (33-year-old female). "My daughter likes me to tickle her arm or her back at night to get her to sleep. Tickle comforts her" (33-year-old female). However, medicinal tickle (you'll feel better after . . .), like medicinal sex, is a matter of taste and timing. Ticklers are generally more confident than ticklees of the healing power of their nonreligious "laying on of hands," as documented by a 20-year-old male. "Some people will tickle to 'lighten the mood' or 'cheer me up.' It just makes it worse; now I'm angry and resentful toward the tickler." This student reminds us that we should be cautious about assuming that the smiles and laughter of the ticklee are associated with pleasure and good cheer.

We turn now to the often-noted paradoxical relation between tickle and laughter. Why do we seek and pay for laughter evoked by comedy but not that by tickle? This question is not quite the mystery it first seems. Consider the many already-noted instances where tickle and associated laughter may be actively sought, for example, during sex play with a significant other, or rowdy play with family and friends. Many tickle-related pursuits are more enticing than a night at a comedy club.

But what of the matter of paying to hear comedy and not paying to be tickled? A weakness of this argument is that people sometimes *do* pay to be touched and perhaps tickled. Sexual prostitution is not the only instance. Consider the stimulation of hairdressers, massage therapists, manicurists, and pedicurists. Practitioners of these professions administer socially acceptable touch and perhaps even a variation of tickle for a price. (In pedicure and massage, practitioners have told me that they use "deep pressure" and slow strokes to reduce "undesirable" tickle.)

While people sometimes pay to be touched, and perhaps tickled, will they pay to tickle someone? "Yes!" Tens of thousands of people of all ages bought the Tickle Me Elmo doll. Many people also engage in tickle as sexual foreplay, so some seek and a few may even pay for this touch experience. And one already-mentioned person will even pay to watch.

The effects of tickle are generally benign, but there is potential for harm if creative minds are put to the task. Tickle can cause you to laugh uncontrollably, gasp for breath, and maybe even worse. The Marquis de Sade (*Juliette*), a tireless if tiresome expert about such matters, and Saint-Foix (cited in *Anomalies and Curiosities of Medicine*) reported that during the sixteenth century, the Moravian Brothers, a Protestant sect of Anabaptists opposed to bloodletting, executed transgressors by tickling them to death. Centuries earlier during the persecution of the Albigenses (heretics) during the French religious wars, Simon de Montfort executed some captives by tickling the soles of their feet with a feather. The ultimate cause of death in such cases is unknown, but the sustained, uncontrollable laughter and struggling of the victim may cause cardiac arrest or cerebral hemorrhage. A widely reported but casually documented type of tickle torture involved coating the bottom of the feet of an immobilized victim with salt that was licked off by goats, presumably tickling the victim who laughed himself to death.

Is there an aversion to tickle or touch so great that it interferes with the conduct of life? Consider the following quotes. "From as far back as I can remember, I always hated to be hugged. I wanted to experience the good feeling of being hugged, but it was just too overwhelming." Or "Many . . . children crave pressure stimulation even though they cannot tolerate being touched. It is easier for a person . . . to tolerate touch if he or she initiates it." Both quotes are excerpted from Temple Grandin's remarkable book *Thinking in Pictures,* an account of her life as a high-functioning and professionally successful person with autism (an "autist"). Her descriptions of the response of autists to touch are reminiscent of those of our normal subjects to highly aversive tickle, especially in regard to the social character of the offending stimulus.

The physical quality of touch is not inherently aversive to autists—it is, rather, the touch of other people that is to be avoided. Ironically, autistic people often crave touch if it is self-initiated, sometimes crawling under mattresses or wrapping themselves in blankets in search of desired deep touch. Grandin designed and constructed her own "squeeze machine" to provide soothing pressure—a way of satisfying her need for being held without actually being touched by another person. (Her machine was inspired by a restraining and quieting device used for livestock.)

When touched by others, autists may experience something similar to an intense tickle sensation and the associated defensive reaction experienced by nonautists. Touch aversion in autism may originate in pathology of the ancient self/nonself identification process involved in tickle. Tickle may tap a yet unrecognized dimension of the deterioration of social competence that characterizes the disorder. Support for this proposition comes from studies showing the cerebellum to be abnormal in some autists. The cerebellum is the brain region shown by a functional imaging study (below) to be involved in the processing of tickle information.

Autism is an extraordinary experiment of nature that provides a privileged glimpse into the veiled process of tickle. Although autism is a pervasive developmental disorder (onset before three years of age) that obviously involves more than an aberration of the tickle mechanism (it also typically includes mental retardation, impaired social interaction and language development, and the performance of repetitive, stereotyped movements), it's useful to consider the extent to which such pathology may be an extreme variant or disorder of a process (tickle) possessed by everyone, rather than a unique pathology. The labeling and categorization of pathology may distort our perception of a continuous process. Autism, for example, may involve pathology of the brain process generating the boundaries of "self" and provide insights into the nonself detection system of tickle. The autists who tolerate or actively seek tickle may have a better developed sense of self/nonself than the tickle avoiders.

Tickle is a part of our lives from shortly after birth. Socially stimulated laughter develops around three and a half to four months of life, and tickle (tactilely stimulated laughter) appears about this time or shortly after. Because tickle requires the discrimination of self from nonself, the study of tickle and laugh development may provide indirect evidence about the emergence of self in infants, although at a level very different than that considered by personality theorists. The significance of laughter and tickle to the baby is well established. Babies laugh when tickled/touched by their mothers, a vocalization that leads to more maternal contact and baby laughter up to the point when the baby is overstimulated and fusses, causing the mother to stop. Laughter, smiling, crying, and fussing, whether evoked by tickle or other stimuli, are important means by which preverbal infants control the behavior of their mother and other caregivers. After speech develops, laughter, smiling, tickle, and other nonverbal signals remain important channels of communication with parents, family, friends, and lovers.

Throughout life, the characteristic reciprocity of tickle resembles that described between human and chimpanzee mothers and their babies (Chapter 5). In both species, it serves a similar function, the establishment and maintenance of relationships. My respondents understood tickle's role in the give-and-take of social engagement. "I like to tickle my boyfriend when it will entice him to tickle me back" (21-year-old female). "I liked tickling my brother to get him to chase me and play tag when we were young" (21-year-old female). Reciprocity was essential to the pleasurable experience of one 22-year-old male tickler. "I dislike tickling someone who doesn't laugh, struggle, or try to get me back." Sometimes the reciprocity of the ticklefest takes on the language of retribution. "I like to tickle someone only to get them back when they are tickling me" (20-year-old female). And "I don't like tickling someone when I can't get away without retaliation" (21-year-old female). The desire for reciprocity/retaliation sustains "tickle battles," perhaps the most benign form of human conflict. In the give-and-take of tickle combat, we see human primates at their most chimplike.

As we age, however, tickle loses much of its attractiveness. I was reminded of this by a distinguished-looking elderly woman who volunteered to fill out one of my tickle questionnaires after a banquet at which I was the guest speaker. She said "I'm an 83-year-old woman, and you want me to answer questions like these?" My immediate concern that she was offended faded as she explained that such questions about tickling and being tickled were simply not relevant to the life of an elderly woman. I reminded her that I was interested in her response for exactly that reason—to understand the tickle experiences of the elderly, whatever they may be. The observation of my elderly respondent was borne out by our questionnaire results. As we age, we tickle and are tickled less, our ticklishness declines, and our pleasure from tickling and being tickled decreases. I was unprepared, however, for the magnitude and early onset of these effects.

Startling changes in tickle are obvious by middle age, a stage of life not usually associated with marked physical decline. I stumbled upon one of life's transitions previously unappreciated by celebrants of their fortieth year. Mid-life involves a gradual tactile disengagement, whether by choice or social circumstance. These changes are not trivial—they involve major declines in the mammalian triad of tickle, touch, and play, related behaviors at the root of our social and emotional being.

Respondents younger than 40 years old were more than 10 times as likely as those 40 and over to report having been tickled during the past week (43 percent versus 4 percent). An even greater proportion of the younger group reported having tickled someone during the previous week (47 percent versus 1 percent). Mid-life marks the reduction in tickle games and relationships, a change in how we relate to people, and how they relate to us. Although the fact of this transition is well established, its cause is less clear.

Several factors implicate endogenous, possibly hormonal factors in the age-related decline in tickle. When the over-40 age group was tickled, they liked it less than younger subjects, giving it a pleasure rating of only 3.5, versus 5.3 (on a 10-point scale) for subjects 39 years and younger. The older group's relative dislike of being tickled may be associated with their lower self-reported rating of ticklishness compared with the younger group (4.7 versus 6.8 on a scale ranging from 1, not ticklish, to 10, very ticklish). The over-40 group also enjoyed tickling others less than the younger group (4.0 versus 6.2).

The social arena of tickle also contracts during middle-age—courting and associated sex frolics decline, and there are fewer chances for physical intimacy with adult children who are moving out of the household. The age structure of tickle relationships also contributes to tickle's age-related decline.

When was the last time you yearned to tickle an old person? Or to be tickled by one? My respondents showed little such interest. Only 3 percent thought it "most likely" that they would tickle someone "older than me." Most said they were most likely to tickle someone "younger than me" or "about my age." Thus, with increasing age, there is a decreasing pool of potential ticklers who will engage you in a tickle relationship, with grandparenthood offering a last hurrah.

Tickle has a clear developmental course. The nonspecific arousal of children in rowdy play gives way to the heterosexual sex play of adolescence and early adulthood, and the tickle games bonding parents and children. A 49-year-old female nicely captured this maturational sequence. "I think it changes as you age. In childhood it's fun . . . but can quickly turn into a control issue; this is especially true when adults tickle a child to tears. As sensuality and sexuality develops, I believe that becomes the primary goal." (This woman reported being tickled "to the point of orgasm.") Tickle gradually fades after the parenting years of middle adulthood, dimming further during old age, periodically being rekindled by the tickling of grandchildren and other children. Tickle and touch also linger surreptitiously in the forementioned hired caresses of hairdressers, pedicurists, and manicurists. But few older people lust for the tickle of rowdy youth—most give tickle and tickling a low pleasure rating.

Tickle is a strange behavior, but we need not search for an exotic neural mechanism to explain it. A well-known neural process accounts for many of tickle's perplexing qualities. The central clue about the nature of the tickle mechanism is this: We can't tickle ourselves. Over 2,000 years ago Aristotle showed acute intuition about this phenomenon: "Is it because one also feels tickling by another person less if one knows beforehand that it is going to take place, and more if one does not foresee it? A man will therefore feel tickling least when he is causing it and knows that he is doing so." Lawrence Weiskrantz and his colleagues devised a manually operated tickle machine to explore why you can't tickle yourself and put Aristotle's proposal to the test. Their machine permitted three levels of subject control over a stroking stimulus applied to the sole of a subject's bare foot. The intensity of the tickle sensation varied inversely with subjects' control over and predictability of the touch stimulus. In other words, the machine tickles most if it moves autonomously or if it zigs when you zag. (The light tickle evoked by the machine produces an urge to scratch or rub the site of stimulation or withdraw the stimulated part, not the more potent sensation that triggers squirming and hardy guffaws.) A human's inability to tickle herself is due to a neurological cancellation of the sensory consequence of self-produced movements.

Weiskrantz's results are consistent with the present social perspective—they are explainable by the action of a *nonself detector*, a neurological comparator that distinguishes self-stimulation of our body (proprioception) from that produced by motile, external objects, or organisms (exteroception). The less predictable the stimulus, the greater its nonselfness and otherness. A nonself, animate stimulus meets the criterion for other, the lowest level of "social" stimulus.

Sarah-Jayne Blakemore and colleagues confirmed and extended Weiskrantz's results using a high-tech descendent of his simple device. Using a robotic tickler controlled by a computer, Blakemore was able to precisely and systematically vary the correlation between joy stick movements produced by a subject's left hand and a corresponding tactile stimulus applied by the robot to the subject's right hand. The subjects' ratings of "tickliness" rose with increasing delay (up to ⅕ second) and trajectory perturbations (up to 90 degrees) between sinusoidal movements produced by the subjects' left hand and the sensory stimulation of the right. The stimulus cancellation that prevents self-stimulation is minimal with zero delay or trajectory perturbation and increases up to a point (⅕ second or 90 degrees) when the sensation becomes indistinguishable from an externally produced sensation.

Using functional MRI (Magnetic Resonance Imaging) technology, Blakemore and colleagues implicated the cerebellum as the site of this stimulus cancellation process. (Functional MRI is a state-of-the-art, noninvasive technique that provides high resolution images of both brain structure and activity.) Of all brain regions, only the cerebellum showed the selective lowering of neural responses to tactile autostimulation relative to external stimulation, evidence of a cancellation process. Tactile autostimulation may trigger cerebellar signals (corollary discharges) that block activation of the somatosensory cortex and the experience of self-tickle and other self-produced tactile stimuli. (As suggested above, a disorder of this system may be implicated in autism.)

Not all tickle science requires robots, computer interfaces, and MRIs. A revelation about tickle came to me recently while I was soaping my foot in the shower. I'm always vigilant for phenomena that confirm or disconfirm the status quo, and this was such a moment. I was startled to find that my stroking the sole of my foot tickled more than it should have given my pronouncements about one's inability to tickle oneself. The ticklishness was particularly strong when I tickled my left foot with my right hand or my right foot with my left hand. The sensation of tickle was decidedly less when I tickled my right foot with my right hand and vice versa. To translate my experience into more general anatomical terms, contralateral (opposite side) tickle had a greater effect than ipsilateral (same side) tickle.

Two days later I pursued these tickle phenomena without the shower and soap, using the 25 right-handed students in my Sensation and Perception class. (The four left-handed students were insufficient to pursue the variable of handedness.) The students were instructed to remove their shoes and socks and sit in a typical cross-legged manner in their seats with the ankle of one leg resting on the knee of the other. They then strummed the sole of their foot with their finger tips, first with one hand and then with the other, rating the relative ticklishness of each foot stimulation on a scale from 10 (very ticklish) to 1 (not ticklish). They then switched legs and tickled their other foot. (The procedure was counterbalanced for foot and hand order.) The students confirmed my shower experience, finding contralateral (opposite side) stimulation to tickle more (4.2 rating) than ipsilateral (same side) stimulation (2.9 rating). On 43 of the 50 trials of stimulation (once for each foot of the 25 students), contralateral tickle was rated as stronger than ipsilateral tickle. (The most potent self-tickle was when the left foot was stimulated by the right hand.)

What does this preliminary, low-tech adventure into tickle science tell us? The key finding is that it tickles more when we reach across the body midline and stimulate the opposite foot. This conclusion makes sense within the context of nonself (otherness) detection considered above. Relative to ipsilateral stimulation, our brain is less likely to recognize contralateral stimulation as self-produced. With ipsilateral tickle, proprioceptive information from the tickling hand and exteroceptive information from the tickled foot enter the spinal cord and ascend on the same side of the body, cross the body mid-line once and arrive at the hypothetical neurological comparator in the brain at roughly the same time. (All neural pathways between the spinal cord and the brain cross the mid-line once—

information from the right side of the body projects to the left side of the brain and vice versa.) With contralateral tickle, information from the tickling hand and the tickled foot arrive at relatively different times because they ascend on different sides of the spinal cord and must cross the body mid-line an additional time to reach the comparator. The synaptic delay and pathway length associated with this extra mid-line crossing increases the disparity in arrival times of contralateral relative to ipsilateral information. As demonstrated by Blakemore, the brain interprets greater arrival time disparity as more otherness, and generates a more intense sensation of tickle. (Blakemore considered only contralateral tickle of the hand and did not contrast ipsilateral with contralateral stimulation.) Although it may seem odd that one side of our body treats the other as foreign, during the first months of life, body-left and body-right function in relative independence, slowly becoming coordinated as the "commissural" nerve fibers that couple the two sides develop. These tickle results remind us that even in adulthood our two halves coexist in harmony but not perfect synchrony.

Christine Harris and Nicholas Christenfeld extended the investigation of the tickle stimulus into the social domain by asking "Can a machine tickle?" They reasoned that if a machine can trigger ticklish laughter, intrapersonal variables could not be a necessary part of the tickle stimulus. So can a machine tickle you? The answer is "Yes!" In contrast to the light tickle (knismesis) just examined by Weiskrantz, Blakemore, and this author, their experimental participants produced a full-blown tickle response complete with squirming and laughter whether they *believed* they were being tickled by a person or a machine. Participants had the soles of their naked feet tickled twice—once by a human experimenter, and once by what they thought was the robotic hand of an automated "tickle machine" displayed in the lab. Participants in the "machine tickle" condition were unaware that their soles were really being manually strummed by a hidden experimenter, not the whirring and vibrating, but otherwise immobile machine. Their ruse was successful. Harris and Christenfeld demonstrated that an intense response by the ticklee does not require a human tickler—their human pseudomachine was sufficient. The potency of the tickle stimulus transcends context, supporting Francis Bacon's (1677) observation that, when tickled, "men even in a grieved state of mind . . . cannot sometimes forbear laughing."

But do these results indicate that tickle is completely free of social context? Not according to the definition of sociality offered in this chapter. Even the impersonal tickle machine provides an already noted low level of social stimulation (e.g., nonself, motile entity), and there is the overwhelming evidence that tickle in everyday life is administered by friends, relatives, and lovers. Interpersonal context may not be necessary for tickle, but it's certainly characteristic.

Although Harris and Christenfeld report some reflexlike properties of tickle (e.g., compelling nature of the stimulus), is tickle a reflex? Not by the standards of the simple and circumscribed knee-jerk or Achilles tendon reflexes. With the presumption of a reflex comes the yet unfulfilled obligation of defining its stimulus, mechanism, and the nature of its response. Classical reflexes, for example, usually have simple, short duration stimuli (e.g., tendon stretch) and responses (e.g., muscle twitch), short response latencies, and responses that are proportionate to the amplitude of the stimuli. The difficulty of meeting these criteria is a reason why reflexes are disappearing as generic explanatory mechanisms in modern behavioral science. The social and physical complexity of the tickle stimulus and the complexity of ticklee behavior argue against the tickle response being a simple reflex. The defensive movements of the ticklee, for example, are complex, variable, goal-directed, and socially motivated. Ticklees may variously hit, kick, or wriggle to rid themselves of the stimulus. Consider also the active involvement of the ticklee who must consciously "do something" about the tickle stimulus, whereas the knee-jerk "just happens." The neurologically programmed reciprocity of tickle locks the ticklee and tickler into the give-and-take that is the basis of play. The tickle response with its laughing/smiling/fending away/huddling/squirming may be better described as higher-order behavior of the sort ethologists term "fixed," "modal" or "stereotyped action patterns" (see Chapter 1 for a discussion of laughter as such behavior), but tickle is not rigidly stereotyped and does not fit neatly into any category.

As all engineers know, designing and building a machine clears away all of the nonsense and gets down to the bare essence of a problem. The challenge of building an effective tickling machine (I've tried) is in creating a synthetic organism, a device that tricks our nervous system into responding to it as if it's a living thing, not an appliance. Engineers working with the more mature technology of bedroom toys (e.g., vibrators) face a related problem. Already we have learned that tickle machines are difficult to construct. The devices of Weiskrantz and Blakemore produced only a light tickliness—the only vigorous ticklish laughter was stimulated by the human pseudomachine of Harris and Christenfeld. An understanding of the critical issues, whether mechanical or biological, are best understood from the perspective of the ticklee.

The basic mechanism of tickle is the nonself detector that operates by subtracting out stimuli produced by our own movements, leaving only those having an external, unpredictable origin that we interpret as tickle. Without such stimulus cancellation, we would be constantly tickling ourselves by accident—the world would be filled with goosey people lurching their way through life in a chain reaction of tactile false alarms. But this cancellation process isn't perfect—if it was, we would experience anaesthesia during self-produced movements.

Consider the details of the mechanism that prevents self-tickling. When we intend to touch our own skin, our brain sends a command *(efferent)* to our muscles to touch our body at a specific place. This motor act of self-touching triggers a corresponding sensory event *(afferent)* from the touched skin that is sent back to the brain where it is compared with a copy *(efferent copy)* of the original outgoing command to the muscles. If the sensory message from the skin matches up with the efferent copy of the command to the muscles, the two cancel each other out and no tickle response is initiated. If your skin is touched by someone or something else, however, there is no efferent copy to cancel the sensory message from the skin, and the touch stimulus is interpreted as exteroceptive (nonself-produced) touch, or especially if it is moving, tickle.

A related process cancels the apparent movement of visual objects produced by your own eye movements, such as those used to scan this text. Because of this cancellation, you are never confused whether your eye or the book is moving, although the image shifts produced by both are identical at the level of the retinal stimulus. Another cancellation process prevents you from becoming motion sick while you are walking—seasickness is produced by similar, but uncancelled, rhythmic stimuli aboard ship.

The detection and removal of nonself moving objects on our body surface (parasites, predators, and aggressors) are crucial, and the tickle response is part of this process. Vulnerable body areas are the most ticklish and are the most vigorously defended. A graph of body regions most likely to be associated with laughter when stimulated was produced by Christine Harris (see Figure 6.1). The four most ticklish regions, in declining order, were the underarms, waist, ribs, and feet. Although informative, the graph is incomplete, omitting many of the body's most ticklish parts. The eyelashes and the hairs in the nostrils and external ear openings trigger intense tickle and a vigorous response when touched, and are important parts of our orifice defense system. Ethical constraints limit exploration of the highly ticklish genital and anal regions, breasts and nipples, all of which are erogenous zones.

Imagine your strong defensive reaction to a scorpion crawling up the side of your neck, or one wriggling its way around your crotch. The heritage of tickle as a defensive reaction is suggested further by the tendency of even pleasurable tickle to trigger squirming, withdrawal of the tickled appendage, huddling, and a defensive, fending-away reaction. The defensive response to a tickling stimulus is goal-directed, involving a neurobehavioral *servomechanism* (feedback-controlled process) dedicated to turning itself off by removing

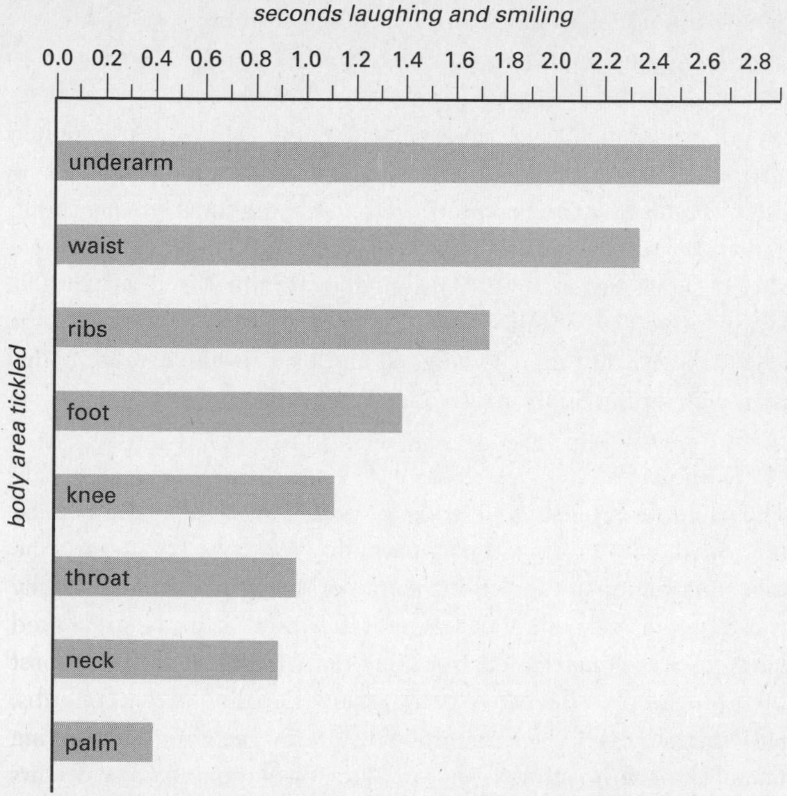

Figure 6.1 Amounts of laughter and smiling produced by tickling different body parts. Of the body regions stimulated, the underarm was the most ticklish. Although the sensation of light tickle can be produced anywhere on the body, only certain "ticklish" areas are capable of triggering laughter. (Adapted from Harris, 1999)

the source of stimulation. (The household thermostat is a servomechanism.) The defensive response is maintained as long as the stimulus is active. The near universal displeasure, kicking, and struggling associated with being restrained and tickled is due to the thwarting of this defensive reaction. "When people tickle my feet, my legs go crazy and I wind up hurting someone. I've just kicked someone in the face" (21-year-old female). A rather similar emergency response is triggered by someone suffocating you by holding their hand over your nose and mouth. You do what is necessary to remove the stimulus, quickly escalating the violence of your reaction. Even infants try to remove such stimuli. The vigorous, defensive reaction to tickle can have tragic consequences as detailed in the civil suit *Sullivan versus Washington Terminal Company.* Sullivan and Garrison were fellow employees working in the glass enclosed signal tower of the Washington Terminal railroad yard in Washington, D.C. While engaged in horseplay, Garrison tickled the very ticklish Sullivan, who laughed and backed into a window that gave way, letting Sullivan fall to his death.

The evolutionary history of tickle, physical play, and laughter are intertwined, with tickle and play predating laughter. Laughter began as a ritualization of the panting sound of rowdy play of which tickle was a trigger and central component (Chapter 5). In the great apes, laughter was emancipated from its original context in the labored breathing of play, the heavy panting now signaling playful intent or anticipation, even when the ongoing level of activity does not demand labored breathing. The laughter of chimpanzees and other nonhuman great apes maintains its ancient pantlike character and association with physical play. Human laughter is a further ritualization (second-order ritualization) in which "ha-ha" is one step removed from the ape pant-laugh and is elicited by a wider range of stimuli, including conversation and the consciously produced and task-oriented cognitive contrivance of humor. Whether the "pant-pant" of apes or the more abstract "ha-ha" of humans, the acoustic structure of laughter is rooted in the respiratory sounds of physical play. *There is nothing arbitrary about the sound of laughter in humans and other great apes.*

The path between tickle and humor is more tortuous and uncertain. Darwin and Hecker advanced the rather implausible idea that tickle and humor were somehow linked because both triggered laughter and had similar qualities as stimuli—joking tickles the mind, etc. The phrase "tickles your funny bone" is a popular expression of a tickle/laughter/comedy relation. Alan Fridlund and Jennifer Loftis found indirect support for this proposition in the strong positive relation between self-reported ticklishness and the tendency to giggle, laugh, and smile. But Harris and Christenfeld counter with an experimental study. Exposing people to humor did not make them more ticklish; neither did tickling make them more susceptible to humor. It's hardly surprising that no link was found between humor and tickle. Darwin and Hecker erred in associating the primitive mechanism of tickle (nonself detection) with the more recently evolved higher-order cognitive process of comedy. The phylogenetically old and new are coupled in other ways. I propose instead the more conservative hypothesis that laughter evoked by jokes is the vocal acknowledgment of verbal play, a response several steps removed from the primal stimulus of tickle.

Upon the ancient nonself touch detection system is superimposed other modern, higher-level brain processes responsible for the many social and emotional nuances of tickle considered in this chapter. The complexity of tickle is enhanced further by its conditionability, increasing the range of stimuli that can trigger it. For example, research by Bobby Newman and his colleagues shows that after a neutral verbal stimulus is paired with tickling, the sound alone can trigger laughs and smiles. Several of my respondents offer anecdotal evidence of conditionability, observing that *simply the threat to tickle can serve as a conditioned stimulus, triggering the conditioned response of squirming and laughter in the ticklee.* "My father would be able to just wiggle his finger at me . . . and I would burst into laughter" (32-year-old female). The power of the "I'm going to get you" game to trigger laughter is based on such a conditioned response.

Can you tickle your dog or cat? The answer is probably yes, although we may have difficulty recognizing the associated behavior because of the lack of recognizable laughter and the impossibility of understanding the subjective experience of these creatures. The flick of a cat's ear or the slap of a cow's tail to shoo flies are responses to tickle stimuli, without a play vocalization. Outside my office window, two young squirrels are at play—it's easy to imagine that reciprocal tickle stimulates and binds them during their lively wrestling and racing through the tree branches. Their rowdy, tumbling frolics resemble human "tickle battles," including their tendency to sometimes conclude in sex. Young dogs, cats, and rats engage in similar escapades. Such rowdy play even scales up to include the largest land animal. Kenyan elephant expert Joyce Poole has seen piles of writhing pachyderms engaging in a bit of rough-and-tumble. But so far, no one has had the strength or inclination to dive into the heap and properly tickle one of these beasts.

Psychologist Jaak Panksepp wisely avoided cavorting elephants in his study of animal tickle, turning instead to his familiar laboratory rats. In the first systematic study of nonhuman tickle, Panksepp's rats responded to his light finger strokes of their ribs and belly with playful nips and ultrasonic "play vocalizations" that may be the rat's equivalent to human laughter. These chirps (50 kHz), about two and a half times higher in pitch than can be heard by the human ear, seemed to signal a readiness for friendly social encounter and were associated with playfulness in untickled rats—the rats that chirped most, played most. Panksepp knows how to pleasure his rats—they treated his hand as a playmate, struggling during tickle, but returning for more. Although Panksepp would be surprised if his rats had a sense of humor, he suggests that they certainly do appear to have a sense of fun. His pioneering work has rewards even for those indifferent to rat recreation and social welfare. It may offer a readily available animal model for the study of positive emotions. The neurochemistry of chirping may lead to the development of new classes of antidepressants and provide insights into the physiology of play and joy.

Tickle may be at the root of all play, triggering generalized arousal in the young, and sexual excitation in adults. Indeed, tickle is one of the reasons that play is fun. The proposal that ticklelike behavior in some form is present in squirrels, dogs, cats, rats, and elsewhere in the animal kingdom is not a foolhardy foray into the murky depths of animal consciousness. It's based on the conservative and parsimonious assumption that so essential a sensory capacity as nonself detection and the associated defense of the body surface must be widespread in the animal kingdom, perhaps even being shared at some level by insects and other invertebrates. However, the sensation of tickle as humans experience it, complete with social and emotional components and a play vocalization, is probably present only in mammals that have well-developed play and a complex pattern of sexual and social relationships.

SEVEN

Contagious Laughter and the Brain

he saying "Laugh and the world laughs with you" (Ella Wilcox, 1850–1919) suggests one of the most remarkable properties of human laughter—its contagion. When we hear laughter, we tend to laugh in turn, producing a behavioral chain reaction that sweeps through a group, creating a crescendo of jocularity or ridicule. The contagious laugh response is immediate and involuntary, involving the most direct communication possible between people—brain to brain—with our intellect just going along for the ride. Contagious laughter is a compelling display of *Homo sapiens,* the social mammal. It strips away our veneer of culture and language and challenges the shaky hypothesis that we are rational creatures in full conscious control of our behavior.

Contagious laughter provides some extraordinary displays of human group behavior. We first examine a plague of laughter in Central Africa that disrupted schools for several years. Lest you think that such outbreaks are confined to Africa, we turn next to an epidemic of "holy laughter" that started in North America and is sweeping through Christian Pentecostal churches around the world, leaving worshipers "drunk in the spirit." For something closer to home, we consider the laugh tracks of television comedy shows, laugh records, and laugh boxes, technological contrivances that tap the same mechanism that drives the epidemic laughter of the African schoolchildren and the Pentecostal worshipers. From these synchronized vocal outbursts come insights into the neurological roots of human social behavior and speech perception.

Laugh Epidemics

Consider the extraordinary events of the 1962 outbreak of contagious laughter in Tanganyika (now Tanzania). The setting was a mission-run boarding school for girls between 12 and 18 years of age at Kashasha village, about 25 miles from Bukoba, near Lake Victoria. The first symptoms appeared on January 30, when three girls started laughing. The symptoms of laughing, crying, and agitation quickly spread to 95 of the 159 students, forcing the school to close on March 18. The school reopened on May 21 but closed again within a month after 57 pupils were stricken. Individual laugh attacks lasted from minutes to a few hours and recurred up to four times. In a few cases, the symptoms persisted for 16 days. Although temporarily debilitating, the laugh attacks produced no fatalities or permanent aftereffects, but teachers reported students being unable to attend to their lessons for several weeks after a laugh episode. The affected girls were highly agitated and often resisted restraint. None of the teachers (two Europeans and three Africans) were afflicted.

The girls sent home from the closed Kashasha school were agents for the further spread of the laugh epidemic. Within 10 days of the school closing, laughter attacks were reported at Nshamba, home village for several of the Kashasha girls. Two hundred seventeen of the 10,000 Nshamba villagers, mostly young adults of both sexes and schoolchildren, were afflicted.

Another outbreak occurred at Ramashenye girls' middle school on the outskirts of Bukoba, near the home of other Kashasha students. The school closed in mid-June when 48 of the 154 girls were overcome with laughter. Kanyangereka village, 20 miles from Bukoba, was the site of yet another outbreak, with one of the Ramashenye girls being the source of the contagion. This flare-up first involved members of the girl's family (sister, brother, mother-in-law, and sister-in-law), but quickly spread to other villagers and to two nearby boys' schools, both of which were forced to close.

Before finally abating two and a half years later, in June 1964, this plague of laughter spread through villages "like a prairie fire," forcing the temporary closing of more than 14 schools and afflicting about 1,000 people in tribes bordering Lake Victoria in Tanganyika and Uganda. Quarantine of infected villages was the only means of blocking the epidemic's advance. A psychogenic, hysterical, origin of the epidemic was established after excluding alternatives such as toxic reaction and encephalitis.

The epidemic grew in a predictable pattern, first affecting adolescent females at the Christian schools, then spreading to mothers and female relatives but not fathers. No cases involving village headmen, policemen, schoolteachers, or other "better educated or more sophisticated people" were recorded. The laughter spread along the lines of tribal, family, and peer affiliation, with females being maximally affected. The greater the relatedness between the victim and witness of a laugh attack, the more likely the witness would be infected.

To consider the Tanganyikan laugh epidemic as an exotic quirk of an alien culture is to miss the broader implications of the phenomenon. Have not we all experienced a lesser form of the epidemic? Recall your own experience with "fits" of nearly uncontrollable laughter (laughing "jags"). Innocent bystanders are also sucked into this vortex of social biology. Once initiated, laughing jags are difficult to extinguish, a point noted by several television newscasters who have suffered laugh attacks during broadcasts. Heroic effects to stifle such outbursts often make things even worse. The laugh tracks of broadcast comedy shows produce their own mini-epidemics in the name of entertainment. The neural mechanism responsible for laugh epidemics replicates behavior that it detects, producing a behavioral chain reaction. Similar mechanisms are involved in the infectiousness of yawning, and perhaps crying, coughing, and other simple, stereotyped acts that are replicated by group members. The pathology of this replication mechanism may be observed in the echolalic speech and echopraxic movements of Tourette disorder (Chapter 8). But the Tanganyikan laughter outbreak and related contagious behavior indicate that the human susceptibility to social synchronization need not arise from pathology.

History is rich in examples of "mass hysteria" and "mania" in which people are swept up into frenzies of communally synchronized behavior. Notable are the St. Vitus's and St. John's dance manias of the European Middle Ages, and the tarantella, an Italian dance craze thought to be caused by the bite of a spider. Secluded groups and women are especially sensitive to mass hysteria, with convents being sites of some of history's stranger epidemics. One nun in a large French convent started mewing like a cat, triggering a chorus of contagious mewing that swept through the sisters. Eventually, the nuns gathered daily for several hours of communal mewing, a performance that continued until stopped by police who threatened to whip those who continued. Even stranger is the epidemic of biting nuns in the fifteenth century. One nun began biting her companions, triggering an epidemic of mutual biting that engaged all of the sisters in the convent, spreading to other convents and eventually to the mother house in Rome.

News reports remind us that behavioral epidemics are still with us. Reports of mass sickness of schoolchildren on trips and outings are particularly common, with adolescent females being especially vulnerable. Common symptoms include fainting, nausea, dizziness, and respiratory distress. (If a toxic reaction is involved, males and females should be affected equally.) Stressful conditions such as anxiety, heat, sleep deprivation, or strange smells, often set the stage for the mass reaction. Rock concerts provide an artistic venue for contagious behavior of a different sort. The so-called Beatlemania of the 1960s, its predecessors and descendants, is not just the hyperbole of press agents. The sea of rhythmically pulsating humanity at these events is a spectacle that titillates even rock-hating social and biological scientists.

This gallery of social exotica offers an important lesson. Although many of these episodes of contagion are bizarre, it's parsimonious to view them as extreme, sometimes pathological instances of acts that are adaptive at lower levels. A similar argument can be made for phobias (i.e., "irrational" fears of height, snakes, closed spaces), panic, and paranoia, responses that are appropriate if triggered by truly threatening circumstances and expressed at more modest levels. Pathologizing extreme cases of normally distributed behavior creates errors of categorization that artificially partition and distort our thinking. It is, after all, often adaptive for social animals to coordinate their behavior: when one animal in a group is spooked and runs, everyone runs; when a mother becomes anxious in the presence of a stranger, so does her infant. At this level, the group can be viewed as a superorganism, with each individual being a sensory and motor organ of the whole, contributing to the well-being of the group and sharing vicariously in its collective experience.

Holy Laughter

Our mouths were full of laughter and our tongues sang aloud for joy.

—Psalms 126:2

Does laughter bring us nearer to God or signify being filled with the Holy Spirit? Exuberant worship, sometimes featuring laughter, has cycled in and out of favor, having a place in the heritage, if not current practice, of many denominations. Before they adopted the more reserved practice of quiet meditation, the original Quakers of mid-1600s England actually quaked. More specifically, they were described to fall, foam at the mouth, roar and swell in their bellies. The early Quakers also had the disconcerting habit of sometimes going about naked as testimony of their faith. And the Shakers of mid-1700s New England actually shook in religious ecstasy—singing, jumping, and dancing were often parts of their religious services. The early Methodists of John Wesley also did a lot of shaking and quaking.

These fervent and animated practices have not disappeared from religious life—the Pentecostal and related "charismatic" churches that continue such traditions are, in fact, the most rapidly growing Christian denominations. The United States has been a fertile ground for the evolution and growth of such sects, some transplanted and some native, particularly among the economically, socially, and religiously disenfranchised. America's unique religious heritage has its roots not in the urban centers of the power elite, but in frontier camp meetings and revivals with names like Cane Ridge, and in the humble Azusa Street Mission in Los Angeles.

"Laughing revivals" offer an intriguing contemporary manifestation of this "old time religion" that provide fascinating case studies in the power of contagious laughter. In the article "Laughing for the Lord," *Time* magazine (15 August 1994) reports the growing popularity of "laughing revivals" among many groups, including the otherwise reserved Anglicans. This "holy laughter" is the adoption of a practice already established in the more flamboyant services of the Pentecostals. Even the historic Cane Ridge (Kentucky) Revival of 1801 had a "laughing exercise."

The visitation of the Holy Spirit to members of the contemporary congregations is signaled by the spread of laughter through the group, followed in some by falling to the floor, sobbing, shaking, twitching, speaking in tongues, and even roaring. In secular contexts, you have probably heard that someone "laughed till they cried," or were "weak in the knees." These evangelicals are exploiting these powerful psychological and physiological consequences of laughter in their religious services. The contagiousness, perseveration (once started, it's hard to stop), and social bonding of laughter (we laugh with our friends and relatives) all work to enhance the power of the laugh experience in a communal setting. Unlike the more subtle practices of prayer or meditation, worshipers can actually *feel* the physiological changes taking place within their bodies during laughter and assign this effect to the divine.

The immediate and powerful consequences of holy laughter are very appealing to those who are self-selecting for such experiences by attending Pentecostal church services. Holy laughter has much in common with glossolalia (speaking in tongues), the defining characteristic of Pentecostal churches. Both laughter and speaking in tongues signify being filled with the Holy Spirit. Since neither are presumed to be under voluntary control, they are accepted as empirical evidence of the divine—the tongue and voice of the spiritually "filled" are under His control. Following this rationale, laughter would be even better evidence of the divine than "tongues," because laughter is under less voluntary control than speech.

Holy laughter is a particularly useful introduction to the Pentecostal experience for novices—it's a potent but achievable first step that is less likely than speaking in tongues to trigger self-doubt or judgmental murmurs among people in the secular world. Once committed to the faith, however, criticism from without is less of an issue and can actually strengthen bonds between the individual and the group. Some Pentecostals even take pride in being known as "God's peculiar people," an affection not extended to the title "Holy Rollers" coined by outsiders.

The resurgence of "holy laughter" today is centered at the Toronto Airport Christian Fellowship, headed by pastor John Arnott. The Toronto Fellowship was influenced by American-based South African Pentecostal revivalist Rodney Howard-Brown, whose services feature outbursts of laughter. Visitors experiencing holy laughter in Toronto carry the "Toronto blessing" back to their home congregations in the tradition of "apostolic succession," propagating this church-specific laughter in what is becoming known in church circles as the "Toronto wave." The tidal wave of holy laughter is spreading worldwide.

I enjoyed the videotaped worship services of Rodney Howard-Brown, the star and leading proponent of the holy laughter movement. Howard-Brown is the self-proclaimed Holy Ghost bartender, who is serving up the "new wine" of holy laughter. He certainly succeeds in getting many in his congregation "drunk in the Spirit," with many staggering or running around, others falling to the floor thrashing, wailing, laughing, speaking in tongues, and making every conceivable animal sound. The power of these services is sometimes so great that people are overcome outside in the parking lot, where, he jokes, "they should not be mistaken for speed bumps put there by the church."

In Howard-Brown's services, laughter is first prompted by his Christian humor, then occurs spontaneously, building in a contagious response, a "Niagara of laughter." People attend these revivals expecting to laugh and they enthusiastically comply with his urging to do so—"Let it bubble out of your belly like a river of living water." Howard-Brown's belly supplies a torrent of its own. He artfully laughs at his own jokes, and throws numerous bombs of his own very infectious laughter at anyone else's guffaws, a tactic that gooses his receptive congregation into the upper reaches of the laugh scale. Although his comic material is not exceptional, Howard-Brown is a master at milking the contagious laugh effect—the basis of laugh boxes, laugh records, and television laugh tracks. He *is* a spiritual laugh box. No comedian so skillfully exploits the laugh-evoking potency of laughter.

At this point, you may have noticed similarities between "holy laughter" and the Tanganyikan epidemic. The giggling girls sent home after the premature closing of their school in Bukoba carried with them the laugh virus that spread to their mostly female relatives, friends, and neighbors. In a parallel fashion, the Western, evangelical version of the laugh epidemic is wildly infectious, being spread from congregation to congregation through a series of "anointings"—one person can transfer his or her anointing to those at another church in a spiritual chain reaction. Although holy laughter differs from its Tanganyikan counterpart in significant ways—infecting both males and females, acting across a wider range of ages, being welcomed, and having context specificity, it would be justified for an African newspaper to report "Laughter Outbreak in Christian Churches," a story balancing that by the *New York Times* (8 August 1963), "Laughing Malady Puzzle in Africa."

Laugh Tracks: From Ancient Greece Through *I Love Lucy*

We turn now from laugh epidemics, religious and secular, to television laugh tracks, a quirky topic apparently more at home in pop culture than social biology. But the laugh track taps the same weird strain in human nature as the laugh epidemics and ultimately suggests the brain mechanism of contagious laughter and laugh perception. Appropriate to the context of the laugh-track story, our attention shifts from the religious to *TV Guide, Variety,* and *Time.*

Laugh tracks have accompanied most television situation comedies (sitcoms) since 7:00 P.M. EST on September 9, 1950. On that evening *The Hank McCune Show*—a comedy about "a likeable blunderer, a devilish fellow who tries to cut corners only to find himself the sucker"—used the first laugh track to compensate for being filmed without a live audience. *Variety* (13 September 1950) was alert to the one innovation in this "fairly amusing" show aired on NBC—"there are chuckles and yocks dubbed in." Sensing something of interest, the reviewer mused, "Whether this induces a jovial mood in home viewers is still to be determined, but the practice may have unlimited possibilities if it's spread to include canned peals of hilarity, thunderous ovations and gasps of sympathy." The outcome of this experiment is all too familiar to contemporary television viewers. The legacy of the laugh track, or canned laughter, long outlived the few, feeble episodes of *The Hank McCune Show*, the last of which aired on December 2, 1950. Forty-nine years later, *Time* magazine included laugh tracks in their list of The 100 Worst Ideas of the Century, along with prohibition, Muzak, and aerosol cheese (*Time*, 14 June 1999).

From the primal moment in 1950 till the present, laugh tracks have been a ubiquitous and controversial presence in broadcast comedy. Although the broadcast industry would rather not talk about laugh tracks, publications including *Time, Newsweek, The New Yorker, Saturday Review,* and *TV Guide* enthusiastically filled the void, especially from the 1950s through the early 1970s. In recent years, controversy about laugh tracks has subsided, probably reflecting an acceptance of the inevitable. Citizens of the 1990s may have forgotten the Cold War political climate of the 1950s, when the laugh track was born, a time of the McCarthy hearings and concerns about brainwashing, hidden persuaders, subliminal messages (the mythical "buy popcorn" commands buried in movies), and communist infiltration of the media. Laugh tracks were seen as a highly suspect and inappropriate effort to control the television viewing masses. Today, we have shifted our concern with mind control from the Red Menace to violence in the media, family values, and Bill Gates. And the medium that most fires our imagination, hopes, and fears is the computer, not the television or radio of generations past. But the challenge of the laugh track as a scientific problem remains, a subtle clue to a process that lies far beyond the scope of popular culture. Like all good scientific detective stories, the curious matter of the laugh track is much more than it seems at first hearing. What begins as a chapter in broadcast history tracked through old copies of *Variety* and *TV Guide* quickly develops into an adventure in social biology that taps one of our species' ancient responses.

The laugh track is a result of a revolution in the relationship between actors and their audience forced by broadcasting. Theatrical performance, especially comedy, involves a collaboration between actors and their audience. This is not precious academic musing about the theory of theater, but a very practical matter.

A successful comedian must be attentive to audience cues that govern timing—the audience must be given an interval in which to laugh or applaud. Comics in a play or a stand-up routine don't want to throw away a great punch line by reciting it while the audience is still digesting or applauding previous material. Audience feedback also influences pacing and the selection of improvised material—comedians go with what works. Successful comedy cannot be done in a vacuum—many actors have commented on how difficult it is to do comedy in a half-empty or indifferent house. A small audience is likely to be an unresponsive audience.

Let us now back up a few thousand years and review some steps in the development of theatrical performance. History provides the clearest examples of the nature and evolution of the relationship between actors and audiences, an essential element of the laugh-track story.

The Roman emperor Nero (A.D. 37–68), an avid actor, had a reliable technique for gathering a responsive audience for his performance—he ordered five thousand of his soldiers to attend and applaud his efforts for which the wise judges always awarded first prize. (Nero particularly liked rhythmic clapping in the Alexandrian style.) Such incentives can drive powerful contagion effects. Who, after all, wants to pan the performance of an insecure and paranoid emperor who takes his art seriously and has a habit of disposing of adversaries, including his own mother? Nero's approach of bringing his own cheering section to his performances was not original. Hired cheering and jeering existed hundreds of years before his time in the Athenian theater of Dionysus where it was used to influence the audiences and judges of tragedy and comedy contests. The comedic playwright Philemon frequently defeated the superior plays of Menander in the fourth century B.C. by this means. The ancient Romans appreciated a powerful technique when they saw it—they even used audiences to influence courtroom verdicts.

The practice of hiring professional claques (from the French *claquer*, "to clap") was initiated in the French theater in the eighteenth century by actors and writers seeking support for their plays. The claque reached its fullest flower in the fertile French soil, but Berlioz, in his *Evenings in the Orchestra* (1852/1963), referred to them as Romans, alluding to both their heritage (he, too, cites Nero) and their association with the Italian contrivance of opera. Although Verdi claims no formal claques existed in Italy, certainly informal ones, both positive and negative, existed and survive to this day. The first nights of Rossini's *Il Barbiere di Siviglia* (1816) and Puccini's *Madame Butterfly* (1904) were marred by troublemakers. In the mid-1950s, a supporter of soprano Renata Tebaldi threw a bunch of radishes at her rival, Maria Callas. According to Sir Rudolph Bing, Manager of the Met, Callas was so nearsighted she thought they were tea roses. In his wonderfully entertaining and catty *5000 Nights at the Opera*, Sir Rudolph himself admits to using a claqueur to energize the audience of soprano Leonie Rysanek, who was a last-minute replacement in *Macbeth* for the unpredictable Maria Callas, a prototypical prima donna. The claqueur shouted, "Brava Callas!" with the intention of arousing audience support for Rysanek as an underdog.

The claque, whether paid or unpaid, formal or informal, endures in modern form mostly in opera, television, radio, and of course political rallies and religious revivals. In all of these scenarios, whether the audience enthusiasm was genuine, augmented, or totally contrived, an audience was present and had a real-time interaction with the performers. Everything changed with the advent of broadcasting.

Radio broadcasting transformed and depersonalized the relationship between actor and audience. The home-listening audience, far larger than that in the theater, was not physically present at the performance and thus was incapable of interacting with the actors. The information flow was one-way, from actor to the home audience. The radio comedian did, however, usually benefit from interactions with a live studio audience, whose laughter, applause, and ambient noise were broadcast along with the comedy material to radio listeners. In a sense, the home listener was a member of an extended audience and laughed contagiously but did not interact with his remote audience mates in the studio.

A step in the direction of canned laughter was taken in 1922 by vaudeville star Ed Wynn in a live radio broadcast of his trademark skit *The Perfect Fool.* Wynn's delivery was crumbling under the stress of performing to a microphone on an empty stage without the benefit of audience feedback. The announcer quickly assembled an impromptu audience of stagehands, technicians, and other actors to supply Wynn with crucial laughter and applause. Ironically, this first use of artificially augmented laughter in broadcasting was to benefit the performer rather than the audience. Ed Wynn's small, hastily recruited audience of colleagues was, however, still a live audience with which he interacted, a setting not very different from that of the typical radio broadcast.

It didn't take radio producers long to realize the importance of a laughing live audience for their comedy shows. They presumed that the home listeners would laugh along with the live studio audience—a reasonable assumption, within limits. Many comedians, however, took special measures to evoke laughs from their studio audience, such as kicking another comic in the pants and resorting to other gags unavailable to the radio audience, who probably wondered what all the merriment was about. Here we have an effort to enhance studio audience laughter to facilitate a contagious response in radio listeners. Some comics lost their jobs as newly developed rating systems revealed that it was possible to succeed with the studio audience and flop with home listeners.

Television brought a revolution in the relationship between the actor and audience—in many cases, the interactive link was totally broken. The big cameras, bright lights, boom microphones and other bulky technology of early television usually forced a retreat to the studio where shows were performed without audiences, either being broadcast live or filmed for later showing. (The nonbroadcast medium of cinema also severs the actor-audience link, but does not typically use laugh tracks—they are not necessary because films are usually viewed in theaters with large responsive audiences in contrast to the many small, socially isolated cells of home viewers hunched before their television sets.)

The technology of the modern television laugh track was primarily the creation of engineer Charlie Douglas, who quit CBS to start his own lucrative and highly secretive business (Northridge Electronics). Douglas's innovation moved beyond the simple recording and playback of actual audience responses by using numerous endless tape loops of prerecorded laughter that could be combined in various combinations and amplitudes to formulate unique "chords" of audience hilarity. The apparatus could be played like a musical instrument by pressing various keys and pedals. One critic noted that some long-dead audience members achieved a strange immortality by having their laughs memorialized in old laugh tracks.

Neither Charlie Douglas nor his numerous clients showed much enthusiasm for publicity about this suspect enterprise. As *TV Guide* (2 July 1966) reported, "mention the name Charlie Douglas and it's like 'Cosa Nostra'—everybody starts whispering. It's the most taboo topic in TV." Who, then, succumbed to the temptations of employing the artificial laugh tracks? In varying degrees and at certain times, most of television's famous and not-so-famous shows used it. *The Burns and Allen Show, The Bob Cummings Show,* and *Ozzie and Harriet* were early devotees. In an interview with *The New Yorker* (10 September 1984), Carroll Pratt, an early trainee of Douglas who later went into business for himself, acknowledges "laughing" shows in the late years of *I Love Lucy, Dennis the Menace, Leave It to Beaver, Father Knows Best, The Beverly Hillbillies, The Mary Tyler Moore Show,* and *M*A*S*H.* (Lucy herself told *TV Guide* [6 November 1953] that she and Desi "are not in favor of 'canned laughter'" and that "'canned laughter' is obviously phony.")

A word should be said in defense of those controversial souls who add the canned laughter, the laugh doctors. Yes, they are often desperately trying to wring some humor out of pathetic scripts, but the overall pattern of their efforts has a basis in reality. As noted earlier (in Chapter 3), most real-life laughter follows ordinary statements—laughter is more about social relationships than jokes. Your life with its own laugh track is like a vast, unending sitcom produced by a very ungifted writer.

Before leaving the subject of laugh tracks, it's only fair to give some more representatives of the television industry a chance to speak for themselves. Producer Alex Gottlieb candidly reported to *Time* (18 February 1957) that canned laughter sounds more authentic than live laughter—"After all, people aren't expert laughers, but the sound effects man is an expert listener." Paramount production chief Doug Creamer explained a lot in an interview with *Newsweek* (12 April 1971), when he confessed to being "terrified at the thought of risking a good show without a laugh-track, because I don't think people know how to react unless you tell them." Jay Sommers, creator/producer of *Green Acres*, said, "People are so conditioned to the laugh track that if they don't hear it they don't know it's a comedy show" (*TV Guide*, 9 July 1966).

The last and least provocative word will go to diplomatic Sylvester (Pat) Weaver, former president of NBC, who is probably best remembered today as the father of actress Sigourney Weaver. He told *Newsweek* (10 January 1955) that "Laughter is a community experience and not an individual one. No one likes to laugh alone, and when you sit in your own living room an honestly made laugh track can project you right into the audience, with the best seat in the house, to enjoy the fun."

Musical Laugh Records

Musical laugh records use recorded laughter to trigger contagious laughter in listeners. Laugh records differ from broadcast laugh tracks in that they offer a musical context instead of jokes or conversation as the raison d'etre for the yuks. Laugh records also come a step closer than laugh tracks to acknowledging the potency of laughter itself to stimulate laughs in listeners. Although these recordings occupy only a small niche in the novelty-music market, some were produced by artists like jazz greats Louis Armstrong, Sidney Bechet, Jelly Roll Morton, and Woody Herman. Legendary musicians didn't mind being silly in their scramble for scarce entertainment dollars.

Given the competition, it's ironic that the first and technically least sophisticated laugh record is the most successful and enduring. "The Okeh Laughing Record" features an anonymous German tavern keeper and his wife, whose highly infectious laughter interrupts an inept trumpet solo. The worldwide success of this record is testimony to the universality of laughter and its infectiousness—it's one of the most popular novelty records of all time. The record was originally produced in Germany shortly after World War I by the Beka label (*Die misglückte Jugendzeit—Original-Lach-Aufnahme*) and was eventually released on Beka's American affiliate Okeh in 1922. "The Laughing Record" was Okeh's hit of the decade, winning out over other Okeh releases by such jazz immortals as Louis Armstrong, King Oliver, and Bix Beiderbecke.

"The Okeh Laughing Record" is still effective 75 years after its release. It's been a reliable laugh-getter in lectures I have given to groups ranging from middle-school students to astronomers at Goddard Space Flight Center. If you obtain your own copy, remember that, as with other laugh-related phenomena, you'll get more laughs if you listen with friends. "The Laughing Record" is still available through Rhino Records, preserver of the good, the bad, and the bizarre of recording history.

The financial if not artistic success of "The Okeh Laughing Record" probably attracted the attention of Louis Armstrong, who tried to cash in with a laugh record of his own, "Laughin' Louis," in 1933. The midsection of this recording features some forced jocularity that interrupts Louis's trumpet playing, a clear attempt to duplicate the format of its Okeh predecessor. While the record features some good trumpet playing, it falls short as a laugh stimulus. Here we see a playful side of Armstrong that dismayed some "serious" jazz fans who thought their living legend should choose his material more carefully. But Louis seems to be having a good time, and the group seems not to have lost precious rehearsal time preparing this footnote in jazz history.

Woody Herman and his orchestra released their entry in the laugh-record Olympics with the "Laughing Boy Blues" of 1938. Early parts of the recording feature the adequate singing of reedman/director Herman and the superimposed, bluesy yuks of Sonny Skylar. Skylar's effort is quite unnatural and not very funny, but is an example of laughter in the service of the melody. He is the clear winner of the weakly contested category of "Best Blues Laugher."

In 1941, the great soprano saxophonist Sidney Bechet and his New Orleans Footwarmers released "Laughing in Rhythm," a title that delivers on its promise. The piece opens with the rhythmic, artificial-sounding, pizzicato laugh notes of band member Henry Goodwin, and moves on to Bechet's saxophone parody of laughter, and later some free-form laughter in the middle section. The singing was provided by trombonist Vic Dickensen. Bechet and his Footwarmers produce a musically satisfying effort that falls short in the humor department. But listening to the wonderful Sidney Bechet is always time well spent, even on this unpromising entry in the novelty-record sweepstakes.

"Hyena Stomp" (1927) by Jelly Roll Morton and his Red Hot Peppers is probably the most complex and interesting example of parodied laughter in jazz recording. "Hyena" is unusual in having laughter as the sole vocalization, although band member and designated laugher Lew Lemar does speak briefly ("That's terrible, Jelly") during a silence between the short introduction and the body of the piece. "Hyena" is notable for its length, rhythmic sophistication, use of rests and pitch, and close replication of the sound of laughter (ho, ha, he, heh, huh) while still remaining a stylized musical vocalization.

"Hyena Stomp" is full of laughter, yet doesn't have laughter in the title, a problem for us laugh detectives trying to track down promising tunes. Conversely, sometimes I was led astray by compositions with laughter in the title but none in the song. The fine John Kirby Orchestra, for example, encourages us to "Keep Smiling, Keep Laughing" (1942), but delivers no laughter. But with some bands, there is always reason to suspect that something goofy is going on, whatever the title.

Spike Jones's "Flight of the Bumblebee" (also known as "The Jones Laughing Record") of 1946 is the only serious contender for a contagious laugh-producing record in the Okeh tradition. Here Spike, the Babe Ruth of funny music, reverts to that old time religion of the Okeh original—the interrupted instrumental soloist, this time trombonist Tommy Pederson. The official story is that Pederson, a first-rate musician, showed up at the studio with a hangover and no music, and unaware of what Spike had in mind. Lying in wait were Frank Leithner, a world-class voluntary sneezer, and several professional laughers. The determined trombonist eventually fell under the spell of his distractors, and his lovely tone and clean double-tonguing started to crumble, all of which are nicely documented in the recording. "The Flight of the Bumblebee" exploits the flustered trombonist and the contagious laughter effect, without a hint of Spike Jones's trademark ensemble work.

Walt Disney's film *Mary Poppins* (starring Julie Andrews) provides the laugh song "I Love to Laugh," sung by Dick Van Dyke, which is based on the theme of contagious laughter. Laughter in the film caused both contagious laughter and levitation among cast members, a most unnatural state of affairs. The more they laughed, the more they levitated.

Laughter occasionally occurs in the lyrics of popular music, although not always with the apparent intent to trigger contagious laughter in listeners. Examples range from Gershwin's wonderful "They All Laughed" (laughter is often added in performance although it's not present in the score) to the downright strange—Dr. Demento's "The Cockroach That Ate Cincinnati" and "They're Coming to Take Me Away, Ha-Haa," The Monkees' "Laugh," and the Beatles' "I Am the Walrus." If this popular music seems bizarre, recall what high-culture offered in the brilliantly musicked, industrial-strength melodrama of opera (Chapter 4).

We close with one of the most artistically satisfying and delightful of all laugh pieces, the air and chorus "Haste Thee, Nymph," from Handel's oratorio *L'Allegro, il Penseroso ed il Moderato* (1740). Although the laughter of this work is purely in the service of the music and not realistic, it triggered contagious laughter in at least one significant audience, according to Michael Kelly, Irish tenor, friend of Mozart, and famous singer of Handel's oratorios. Kelly reminds us that musically triggered contagious laughter worked its magic long before the laugh-records of the recording age.

As Kelly reported, "Mr. Harrison, my predecessor . . . , was a charming singer . . . but in the animated songs of Handel he was deficient. I heard him sing the laughing song, without moving a muscle; and determined, though it was a great risk, to sing it my way, and the effect justified the experiment: instead of singing it with the serious tameness of Harrison, I laughed all through it, as I conceived it ought to be sung, and as must have been the intention of the composer: the infection ran; and their Majesties, and the whole audience, as well as the orchestra, were in a roar of laughter, and a signal was given from the royal box to repeat it, and I sang it again with increased effect."

The contagious laughter of the royal family during its performance helped to make laughing fashionable again, reversing an attitude typified by the wonderfully pompous Lord Chesterfield, a curmudgeon for all seasons, who advised "A man of parts and fashion is therefore only seen to smile, but never heard to laugh." Once again, laughter cycled back into fashion in Merry Old England.

Laugh Boxes

Nowhere is the contagiousness of laughter demonstrated more effectively than with the laugh box. Push a button and these small, battery-powered playback devices provide a burst of prerecorded laughter that makes us laugh in response. We don't decide to laugh—it just happens. Laugh boxes differ from laugh records and laugh tracks in providing only the sound of naked laughter—they are like television laugh tracks stripped of their jokes, or laugh records stripped of their music. These inexpensive gizmos from novelty and magic shops provide a critical demonstration that in the absence of a joke or humorous remark, laughter, by itself, can evoke laughter.

Laugh-box science is like a lot of other laugh science in being a highly democratic affair—get a laugh box and you are in business.

Most cost between $10 and $15 and vary in quality and funniness of the recorded laughter. The popular Tickle Me Elmo doll is a fancy, upscale laugh box, responding to a belly press with laughter, finally vibrating with a simulated tickle response after the third press, a tactile stimulus that triggers its own laugh and smiles. (For our studies it's preferable to have a less gifted and versatile device that simply produces laughter.) The incredible success of this $30 Elmo doll indicates the potency and desirability of the laugh stimulus. (For a tickle perspective of Elmo, see Chapter 6.)

A laugh-box test was conducted in three of my undergraduate college classes. The 19-second burst of canned laughter produced by my laugh box was repeated 10 times, with the beginning of each trial separated by a one-minute interval. Students were asked to record on a note card whether they laughed and/or smiled on each trial. I was a bit nervous the first time I tried this experiment. Students can be a tough crowd, being reluctant to laugh or smile, even if tempted to do so, as a signal of their displeasure with being manipulated. My concern was unwarranted. The laugh box was a hit, a testimony to the power of the contagious laugh response.

On the first trial, nearly half of the 128 students in three classes laughed at the stimulus laughter, while more than 90 percent smiled. However, the effectiveness of the stimulus declined with each repetition, until only 3 of the 128 students laughed on the tenth trial. Laughter eventually lost its magic, taking a darker tone. By trial 10, about 75 percent of the students rated the laugh stimulus as "obnoxious." (The inquiry about obnoxiousness was made only on the last trial.) The ratings weren't really necessary—student grimaces revealed their growing displeasure during the last few painful trials.

Colleagues with offices adjacent to mine can attest to the aversiveness of canned laughter. I certainly wince every time one of the laugh boxes is accidentally activated. By simultaneously activating several laugh boxes—I have 100 of them—I've discovered a particularly unpleasant jeering effect. The technology of jeering has much less commercial appeal than traditional laugh tracks. People will pay to laugh but not to be laughed at, with the possible exception of audiences of insult comedians like Don Rickles. The strong negative effect of repeated laughter goes beyond the response expected from the recurrent exposure to a generic auditory stimulus, such as "Hello, my name is George," "Hello, my name is George," etc. The negative reaction reflects the deep biological significance of laughter, which in this case has a negative connotation.

Neurological Laugh-Detectors and Social Biology

The animal chorus of contagious laughter has its roots in the neurological mechanism of laugh detection and generation. Here we find the common link between the African laugh epidemic and Pentecostal holy laughter, laugh boxes and television laugh tracks, and claques from ancient Greece to the New York Met. Here also is the source of laughter's power that so concerned Plato and Aristotle (Chapter 2).

The ability of laughter to elicit contagious laughter raises the intriguing possibility that humans have an auditory *laugh-detector*—a neural circuit in our brain that responds exclusively to laughter. (Contagious yawning may involve a similar process in the visual domain.) Once triggered, the laugh-detector activates ("releases") a *laugh-generator*, a neural circuit that produces the movement that we hear as laughter.

Previous accounts of contagious laughter by social scientists ignored the neurological basis of the phenomenon, focusing instead on whether audience laughter increased the likelihood that an individual would laugh at a joke or rate it as humorous. (They generally found that it did.) The power of naked laughter to trigger laughter got lost in a blizzard of sometimes baroque theorizing about such higher-order social processes as "deindividuation," "release restraint mediated by imitation," "social facilitation," "emergence of social norms," "conformity," "peer pressure," or "modeling." Although these processes may facilitate group laughter, it seems that social scientists are willing to do a lot of work to avoid acknowledging that laughter has the innate capacity to trigger laughter.

Figure 7.1 Neurobiological mechanism for the detection and replication of laughter. The ability of laughter to stimulate contagious laughter in another individual suggests that humans have an "auditory feature detector," a neurological detector that responds specifically to the sound of laughter. In turn, the feature detector activates the neurological "laugh generator" that produces the stereotyped movements of the thorax, larynx, and vocal track that create the sound of laughter. (From Provine, 1996)

Evidence for the proposed neurological laugh-detector and its associated laugh-generator is indirect but compelling. Most of what we know about the laugh-generator is based upon its vocal output, themes of Chapter 4 and Chapter 5. We know, for example, that laughter is innate, stereotyped, and present in all members of our species. These properties are consistent with the presence of a neural generator, whether the movement produced is laughter or walking. The same properties favor the evolution of a neurological detector for laughter. And laughter is the type of important species-typical vocalization for which a detector would be selected by evolutionary processes. A detector is unlikely to evolve for an insignificant, arbitrary and ever-changing vocalization possessed by only a few members of our species.

The search for the laugh-detector offers more than the key to explaining communal laugh-fests. Knowledge of the laugh-detector may aid the search for detectors of the complex and variable phonemic features of speech. On a more general level, the study of contagious laughter suggests an experimental approach to human group behavior that moves seamlessly between the neuronal and social levels of analysis. And the focus on testable propositions about neurological mechanisms escapes the fanciful "just so" evolutionary scenarios of some approaches to social behavior. Humankind is at its most formidable and sometimes terrible when it acts en masse, whether in war against a common foe, in response to natural crisis, or in heroic pursuit of a great ideal. In contagious laughter we have an example of how to examine such disparate threads of human group behavior and bring them into the fold of neuroscience.

EIGHT

Abnormal and Inappropriate Laughter

Clinical Perspectives

Why dost thou laugh? It fits not with this hour.

—Shakespeare, *Titus Andronicus*

It was a curious laugh; distinct, formal, mirthless . . . The laugh was repeated in its low, syllabic tone, and terminated in an odd murmur . . . the laugh was as tragic, as preternatural a laugh as any I ever heard . . . and, but that it was high noon, and that no circumstance of ghostliness accompanied the curious cachination . . . I should have been superstitiously afraid.

—Charlotte Brontë, *Jane Eyre*

Titus Andronicus, the Roman general of the quoted Shakespearian tragedy, laughed upon the return of his own severed hand and the heads of his two sons whose lives his hand was supposed to ransom. Such inappropriate laughter is not solely the product of the playwright's fancy. People in tragic circumstances have laughed themselves to death, and sometimes perfectly normal laughter occurs in completely unsuitable social contexts such as funerals, often to the horror of the laugher. In certain rare cases, bursts of laughter may occur independently of social context, appropriate or otherwise, as when driven by the electrical storms of the epileptic brain. And the structure of laughter itself can be abnormal, as with the "curious," "preternatural" laughter (later "demoniac—low, suppressed, and deep") that unnerved heroine Jane Eyre in the quote from the Charlotte Brontë novel. And what of the properties of laughter reported in earlier chapters, those involving the punctuation of speech, speaker/audience balance, acoustic structure, conscious control and contagion? Each of these new features reveals an associated pathology, once we start to look for it. This chapter attempts to bring order to a rogues' gallery of misfortune, a mix of the unusual, the improper, and the bizarre, first by cataloguing, and then by providing an inventory for the description, analysis, and comparison of the types of pathological laughter.

This survey of pathological laughter is full of twists and turns, cobbled together from a small research literature, obscure clinical notes, and miscellaneous odds and ends. It's made of found parts that sometimes don't quite fit, but care was taken not to patch over the gaps to provide a more esthetically pleasing, if less accurate, whole. This potential source of frustration for the scholar and storyteller signals opportunity for the enterprising scientist and clinician, for it suggests critical research yet to be done. Indeed, the stakes are higher than they seem, as inappropriate laughter and crying are among the most common and least understood symptoms of neuropathology and psychopathology.

This foray to the fringes of human experience stresses laughter and its social settings. Until more is known about laughter, behavioral analyses offer the most rigorous approach to the mechanisms of mirth and their pathology. For those desiring a more biological account, I offer that behavior *is* physiology and defines the neurological processes being studied. This conservative approach to behavioral neurology avoids the fanciful circuit diagrams that infect some of the clinical literature and the phrenological folly often celebrated in the popular press. Although pathology indicates that some parts of the brain are more involved than others in laughter and humor, we should be wary of reports that a single brain center/gene/ neurotransmitter is responsible for these complex and distributed processes.

"The Laughing Death" is the headline of a *Time* magazine (11 November 1957) story announcing the discovery of kuru, a degenerative disease of the central nervous system, among the isolated Fore people of the New Guinea highlands. Of special interest to us here is the hilarity that accompanies the early stages of this always fatal disease. But the kuru saga is also full of sorcery, cannibalism, and ritual killing, the exotica of anthropologists' dreams. Add to this potent brew some breakthrough medical science that eventually wins a Nobel Prize for physician/anthropologist/virologist D. Carleton Gajdusek, and you have material as at home in the *New England Journal of Medicine* as in the *National Enquirer.*

Kuru ("shivers" or "trembles" in the Foreans' native language) principally affects adult females, and was the leading cause of death among the Fore during the late 1950s and early 1960s. But kuru is a doubly deadly disease. Because the natives believed kuru was caused by sorcery, relatives of the immediate victim often performed ritual, vendetta killing (Tukabu) of the suspected (always male) sorcerer presumed to have worked his deadly magic. (The sorcerer's limb muscles, kidney region, and genitals were pulverized with stones, and his trachea was crushed by biting.) The kuru story gains momentum because the enigmatic viruslike infectious agent, a self-replicating protein called a prion, is transmitted by the ritual cannibalism of dead relatives. Women, the principal kuru victims, were most involved in the preparation of the bodies for cooking, and consumed the brain, the most infectious human flesh. The very long (4- to 26-year) incubation period of the infectious "slow virus" of kuru prevented the near Stone Age Fore villagers from associating cannibalism with the disease, and likewise complicated scientific study—for several years the sorcery hypothesis fared about as well as those of geneticists, toxicologists, nutritionists, and virologists.

Kuru is nearing extinction as the cannibalism that sustained it, and its last victims, are disappearing. But threats from kurulike transmissible spongiform encephalopathies (TSEs) have not been eradicated. Similar infectious agents are responsible for the rare but always fatal Creutzfeldt-Jakob disease in humans, scrapie in sheep, and bovine spongiform encephalopathy ("mad cow disease"). As Richard Rhodes reminds us in his *Deadly Feasts,* there are many good reasons not to eat our dead, and perhaps even the dead of other animals. Even ordering your meat "well-done" does not protect you from the remarkably heat-resistant mad-cow virus.

During early stages of kuru, victims often show "marked emotionalism" with "inappropriate excessive laughter," or "paroxysmal hilarity," often "laughing at their own stumbling gait and falls," with their kinsfolk joining in. The sketchy clinical reports of kuru laughter suggest that it is "felt" (e.g., associated with euphoria), normal in structure and social context, and is socially effective as measured by its contagiousness, but is more frequent and robust than normal, and may occur in cognitively inappropriate situations. The laughter is not a vocal tic associated with a seizure state. As kuru progresses, increasingly strong tremors and uncontrollable voluntary and involuntary movements render the victim immobile, incontinent, unable to sit, eat, and communicate, although the intellect and consciousness are relatively unimpaired until near the end. The misplaced euphoria of early stages may provide a measure of comfort in meeting this grim and certain fate. The relentlessly progressive disease usually kills victims within a year of symptom onset, typically by causes secondary to the chronic condition (e.g., bronchopneumonia, gangrene).

It's difficult to associate kuru laughter with a particular brain site because the disease causes widespread degeneration and loss of brain neurons, producing the characteristic spongelike (spongiform) state, especially in the cerebellum, the brain region most involved in movement coordination. Focal brain regions implicated in the production and control of laughter by other studies of seizures, tumors, lesions, and stimulation, are certainly affected by kuru, but the excessive laughter and exhilaration may be caused in part by nonspecific excitation and disinhibition. But kuru is not alone in producing symptoms of pathologically enhanced hilarity.

Masque manganique, a jovial but fixed facial expression, and spasmodic laughter signal the grim syndrome of manganese poisoning, a progressive and irreversible neurological disorder described in Moroccan miners. Other symptoms include clumsiness, abnormally strong reflexes, slowness of speech, and hyperemotionalism, with displaced euphoria being much more common than weeping. "When a number of patients are gathered together, this laughing spreads infectiously, mostly evoked by trivialities and quite disproportionate to the events or motions provoking it." Although reports of this toxic syndrome refer to spasmodic laughter, the vocalization has many normal features, including its association with euphoria and smiling, social context, and capacity to evoke contagious laughter, at least in other patients. Little is known about its neural mechanism, but like kuru, it probably involves disinhibition.

The term "sardonic laughter," referring to the bitter, mocking laughter of derision, has a rich if dark etymology. The ancients who coined the term were referring to the humorless laughter and smiling produced by a deadly plant native to Sardinia, probably the herb known variously as marsh (cursed) crowfoot, buttercup, or wild parsley (*Ranunculus sceleratus*). The toxic effect of the plant was well known in ancient times, because the derivative expression had wide early use, as in *The Odyssey,* when Odysseus "smiled in his anger a very sardonic smile." Writing in the second century A.D., Pausanias noted of Sardinia, "The whole island is free of lethal drugs except one weed; the deadly herb looks like celery, but they say if you eat it you die of laughing. That is why Homer and the people of his time speak of something very unhealthy as a Sardonic laugh." Although the details of this mysterious herb and its effects are lost in the fog of history, the term *risus sardonicus*, literally "laughter of Sardinia," but now usually referring only to smiling, survives in modern medicine as a key symptom of tetanus ("lockjaw") and strychnine poisoning. (In antiquity, I suspect that the herb produced only grimaced smiling, as in *The Odyssey*, with "laughter," if any was ever present, the result of rhythmic gasping during seizures.) Both strychnine and the deadly tetanus toxin produced by the bacterium *Clostridium tetani* produce runaway excitation of the nervous system by blocking neural inhibition, like a car careering down a hill after its brakes have failed. The result is the frozen, grinning mask of the sardonic smile, the default condition of the all-out contraction of the muscles of the face and jaw. Other muscles are also involved—in extreme cases of tonic muscular seizures, the victims will have their backs arched in an inverted U-shape, with nothing but the back of the head and tips of the toes in contact with the ground. Death by asphyxiation of the fully conscious victim follows when the seizures paralyze the muscles of breathing, leading to a terrible and grotesque death.

"The atmosphere of the highest of all possible heavens must be composed of this gas," enthused Robert Southey, a view shared by fellow poet Samuel Taylor Coleridge, who under its spell reported "more unmingled pleasure than I had ever before experienced." The gas that so enraptured them is nitrous oxide (N_2O), the dental analgesic and anesthetic known popularly as "laughing gas."

Nitrous oxide was discovered in 1772 by Joseph Priestley, who had discovered oxygen the previous year. In 1799 another patriarch of chemistry, Humphry Davy, as a young man recklessly inhaled the gas, experiencing a most agreeable sense of well-being, relaxation, and giddiness, the responses that earned nitrous oxide its popular name.

In the entertainment-starved United States of the early and mid-

1800s, people enjoyed getting high and watching others do so. Many university students of chemistry and medicine held "laughing gas parties" and "ether frolics," an unpublicized perk of higher education in the sciences. Numerous itinerant, self-proclaimed "professors," "philosophers," and "practical chemists" roamed the towns and crossroads of America giving lectures on chemistry and demonstrating the glories of nitrous oxide or ether (another laugh-producing intoxicant) on volunteers from the audience. Under the influence of the gas, the intoxicated volunteers would laugh uncontrollably, stagger about, and talk foolishly, to the amusement of friends and onlookers.

The popularity of these laughing-gas shows is reflected in ads in a defunct local newspaper, the *Baltimore American and Commercial Advertiser* (later the *Baltimore American*). Seven ads for "laughing gas," or "exhilarating gas" shows were found in the newspaper between 1820 and 1830. (Many less formal sidewalk demonstrations were not advertised.) Among the laughing-gas showmen operating out of Baltimore in the early 1830s was young Samuel Colt ("Dr. Coult"), who later gained fame as inventor of the revolver. By 1844, P. T. Barnum, the quintessential American showman and promoter, got into the business, opening laughing-gas shows at Peale's New York Museum (under Barnum's management) and at his American Museum in New York City. Laughing gas was even a theme of popular music, including "Laughing Gas or a Night at the Polytechnic," by John Nash, and "Laughing Gas" by W. H. Freeman.

Ironically, it was the recreational effect of laughing gas that led to the discovery of anesthesia. While attending a laughing-gas show, dentist Horace Wells observed that an intoxicated volunteer did not flinch in response to what should have been a painful accident. Wells immediately recognized the potential of nitrous oxide for dentistry and had one of his own bothersome wisdom teeth painlessly extracted the next day under the influence of the gas. From this beginning on December 11, 1844, came painless dentistry, and a year later, the use of ether anesthesia in medical surgery by W. T. G. Morton. (While ether produces *anesthesia,* the reversible absence of perception of all senses, nitrous oxide, in contrast, produces *analgesia,* the imperception of pain, but only weak anesthesia.)

The discovery of the anesthetic effects of ether swept away interest in nitrous oxide, but N_2O quickly made a comeback in dentistry and is still used to control the pain and anxiety of dental patients. Some dentists self-administer nitrous oxide, and a few have even died in their chairs, gas masks strapped to their faces, ascending to dental nirvana under the influence of the addictive gas that so inspired poets Coleridge and Southey.

The power of laughing gas is undeniable, but why does laughing gas work and what does it indicate about the mechanism of laughter? Let's begin with the most obvious question: Is nitrous oxide a magic bullet that triggers laughter by scoring a perfect, pharmacologic bull's-eye on the brain's laugh center? Probably not, because the gas does not reliably produce laughter, and the laughter produced is not an uncontrollable, seizurelike bursting forth with "ha-ha-ha," as would be expected if a laugh center was activated. But if nitrous oxide is not a mainline to the "funny bone," how *does* it trigger laughter? This is a difficult question to answer, because no one understands for certain why nitrous oxide or, for that matter, *any* other inhalant anesthetic/analgesic works. Perhaps the N_2O high and analgesia may be responses to a tiny squirt of endorphins, the brain's own opiate supply. While explaining analgesia, the endorphin hypothesis is not entirely satisfactory, because opiates typically produce a mellow high, not laughter. Like other drugs associated with laugh production (e.g., marijuana, hashish, alcohol, ether, LSD), nitrous oxide probably works through some combination of disinhibition and excitation, amplifying the effects of the same social factors that drive everyday laughter. The laugh-induction of nitrous oxide and related agents lies as much in psychology as in pharmacology—they make everything seem funny.

The last word on this subject goes to Gardner Q. Colton, the "laughing gas professor" who introduced nitrous oxide to Horace

Wells in 1844, and later started his own career in painless dentistry. Colton summarized the social and psychological determinants of the response to this remarkable gas in an ad for his performances. "The effect of the gas is to make those who inhale it either laugh, sing, dance, speak, or fight, & etc. . . . according to the leading trait in their character."

The Child's Garden of Grass, a 1960s-era users' handbook, makes a pitch for marijuana as another pharmacological entrant into the Laughter Hall of Fame. The book considers "funniness" as marijuana's most pleasing effect. "There's a little spot in your mind which tells you when you think something is funny and grass expands that little spot until the little spot takes over and everything is funny. Everything. Your friend's teeth are a riot. A simple 'Hello' brings on storms of laughter. And something which is generally funny, like hearing a good joke or watching the Marx Brothers can turn you into a convulsive maniac, writhing in agony and pleading for help. Going out in public in this mood can be a risky act because of the laughing problem, as you can find yourself laughing at people who are not stoned and fail to see what is so amusing. Sometimes they hit you."

Even the hilariously awful 1936 "warning film" *Reefer Madness* got this part of the story right, announcing that marijuana's "first effect is sudden, violent, uncontrollable laughter." More scholarly sources add that intoxication progresses though humor, giggling, and smiling to uninhibited hilarity.

Marijuana intoxication is not just a cultural quirk of the 1960s or the Jazz Age, as we are reminded by Ernest L. Abel's book *Marihuana, The First Twelve Thousand Years.* Compared to marijuana and hashish (another, more potent, hemp plant derivative), nitrous oxide is a recent blip in the historical record. Hashish is mentioned in some of the earliest Arab writing, and Marco Polo reported laughter among stoned sultans during the sixteenth century. Certain

Egyptian cafes of the last century that provided hashish were known as "maschechels," or houses of hallucinating laughter. Poet Charles Baudelaire, a member of the notorious Club des Haschichins in Paris, describes this mirthful exuberance in his 1860 *Poem of Hashish:* "At first there is a certain absurd, irresistible hilarity that overcomes you . . . The most ordinary words, the most trivial ideas, assume a new and bizarre aspect; you are even surprised at having found them so commonplace until now. Incongruous, unpredictable equations and comparisons, endless puns, and comic sketches keep gushing from your brain."

The exhilaration, manic laugh, and wide grin of hashish resembles the reaction to nitrous oxide. However, unlike the rapidly reversible effects of nitrous oxide, the effects of hashish are more sustained and cumulative. Both hashish and nitrous oxide laughter contrast with the weaker, infrequent laughter of opium with its "beatific" smile. As a disquieting conclusion to our analysis of synthetic pleasure, I note that the muscular hashish grin has been likened to the *risus sardonicus* of the dead.

Laughing till you're "weak in the knees" quite literally describes the Pygmies of Central Africa, our planet's most exuberant laughers. In *The Forest People,* Colin Turnbull describes the pygmy as being "not in the least self-conscious about showing his emotions; he likes to laugh until tears come to his eyes and he is too weak to stand. He then sits or lies down on the ground and laughs still louder." And "When pygmies laugh it is hard not to be affected; they hold on to one another as if for support, slap their sides, snap their fingers, and go through all manner of contortions. If something strikes them as particularly funny they will even roll on the ground." These good-natured forest dwellers exhibit an exaggerated case of behavior we have all witnessed to a lesser degree. But sometimes people actually do drop to the ground when laughing, and unlike the fun-loving pygmies, they have no choice in the matter.

Cataplexy is a sudden spell of weakness triggered most often by laughter, but also by anger, and other emotional excitement, or acts. During a cataplectic attack, a person may collapse into a heap of toneless muscles, a completely reversible state, which may last from seconds to 30 minutes. The patient is alert but helpless during the first minute or so of an attack. Milder, more common forms, may be hardly noticeable, involving only sagging eyelids, jaw, or head. Cataplexy is not rare, afflicting at least 70 percent to 80 percent of the 250,000 patients in the United States suffering from narcolepsy, the abnormal, sudden onset of daytime sleepiness. (Among sleep disorders, only narcolepsy causes cataplexy.) Imagine a fisherman being so excited by hooking a fish that he is too weak to reel it in. Finding that we can paralyze some people with a well-placed punch line, or by saying "boo!," gives a perverse sense of power, and a reminder of the quirkiness of our brain and its body.

As a noncataplexic, are you free of laughter's paralytic powers? Not according to researcher S. Overeem and colleagues. In a study of the H-reflex, a neural mechanism regulating muscle tone, they found that laughter almost totally blocked the reflex in both cataplexic *and* normal control subjects. "Going weak with laughter" is not only a figure of speech, it's a physiological fact. Extrapolating from this finding, we should ask whether laughter should be avoided immediately before or during athletic competitions or other demanding physical challenges.

"Happy puppet" accurately describes people whose lives are constrained by the rare, genetically based neurological condition now known as Angelman disorder. Although the "happy puppet" designation has fallen into disfavor because it's considered offensive by some, people with Angelman are indeed puppetlike, with their limbs apparently moved by strings of an unseen puppeteer. The "puppets" seem cheerful, presenting ample laughter and smiles. But on closer examination, something is very wrong with these individuals and their laughter. A mother observed of her afflicted baby son, "He would lie still for some minutes, then suddenly burst into loud laughter for several minutes without apparent reason . . . It is a nice, natural, healthy laughter, resembling the laughter of a more mature child. But I am afraid of this laughter, it makes me uneasy, I cannot rejoice with my baby." This parent later noted of her now seven-and-a-half-year-old son, "I cannot stand this laughter. It starts for no reason; when looking casually at a household object . . . he bursts out into laughter which goes on for 30 to 40 minutes." Other parents are more appreciative of their child's smiling, upbeat nature, for certainly, these families have much more to contend with than simply inappropriate laughter. No person with Angelman disorder has been able to live an unassisted life.

Angelman children are usually severely retarded, hyperactive, seizure-prone, hypotonic (floppy), and perform jerky, repetitive movements (hand flapping, etc.). Their vocal repertoire includes cries and laughter, signals of physical and emotional states, but rarely spoken words. As babies, they may have decreased crying, cooing, and babbling, early signs of their profound deficiency in vocal communication. Although Angelman children are often able to follow simple vocal commands, their vocal competence advances little beyond what is normally expected during the first three to six months of life.

Much remains to be learned about the communication skills of Angelman patients, and laughter may provide a neglected window into the cognitive and emotional state of these largely speechless individuals. For example, are the "happy puppets" really "happy"? Is their laughter different in degree or kind from those without the disorder? Is their laughter paroxysmal (sudden or uncontrollable), as commonly reported, or simply evoked by physical and social stimuli of an atypical sort? Do they laugh as much when in the presence of other people, as when sitting alone in a room? (If their laughter is primarily social, this would indicate a measure of social competence.) Is their sustained laughter unique, or an extreme case of the laughing jags many of us have experienced? Is the ancient social communication of tickle and associated laughter normal in these socially inattentive individuals? Although such significant questions go unanswered in the sketchy research literature, they probably strike the real experts, the parents and family of Angelman children, as naive and uninformed. The vast experience, sensitivity, and insight of parents and family members make them the obvious partners in future research on Angelman and related disorders (e.g., autism, Williams and Rett disorders). And the Angelman children themselves can teach us something important about the nature of laughter—both they and normal prespeech babies provide an opportunity to study laughter independent of speech, the context of most adult laughter.

Angelman disorder is intriguing at the genetic level, a story still unfolding. Males and females are equally affected, unlike the somewhat similar Rett disorder (see below), which afflicts only females. Most Angelman patients lack a piece of one of their two chromosome 15's, the same deletion observed in patients with Prader-Willi disorder, a very different condition associated with mild retardation and extreme obesity. Curiously, the same genetic defect causes two different disorders, the difference being whether the defect is inherited from the mother, causing Angelman disorder, or from the father, causing Prader-Willi disorder. The differential expression of genotype depending on the parent of origin is known as genomic imprinting.

Angelman laughter may be a response to private events in an experiential world we can never fully comprehend, an exaggeration of the normal, or as suggested by some neurologists, a paroxysmal vocalization that is beyond the patient's control. But there is no evidence that Angelman laughter is driven by brain seizures.

Seizure-produced laughter is well established in an unusual type of epilepsy, a case of which was reported in the journal *Aviation, Space, and Environmental Medicine.* A student naval aviator was flying in formation as a check-out before being declared competent for solo flight. All went well until he began to laugh uproariously, and his plane drifted dangerously close to another aircraft, forcing his instructor to seize the controls to avoid a mid-air collision. This near miss marked the end of the student's aviation career and the beginning of his neurological workup. Inquiries revealed that his symptoms began about 18 months earlier, when he laughed so loudly during sleep that he woke his housemates and sometimes even himself. A videotaped record of one of these nocturnal seizures revealed the stereotypic character of the seizure episodes. He first smiled, then rolled from side to side laughing, raised and scissored his right leg over his left, assumed a fetal position, adjusted his pillow, and returned to sleep. The uncontrollable laughter and associated smiles eventually intruded into his waking hours, including outbursts during officers' meetings. Although the young man couldn't remember the actual laughter, he recalled that the laugh episodes lasted about 10 seconds, were preceded by impaired concentration and an acute sense of hearing, and produced no sense of mirth or merriment. A laughing seizure observed during wakefulness, in the course of getting a neurological exam, was accompanied by rhythmic extensions of his legs and abnormal seizure-related (ictal) brainwaves. His seizures were controlled by an anticonvulsant (phenytoin), although he continued to experience occasional auras associated with seizure onset.

The pilot's symptoms are characteristic of gelastic (laughing) epilepsy, a rare disorder presumed to result, like other forms of epilepsy, from the abnormal, massive synchronized discharge of neurons in the brain. The gelastic seizure experience is like having an unseen alien hijack your body, causing it to bark out an emotionless "ha-ha-ha," a peculiar act that is not answered by laughter from witnesses, only discomfort and looks of amazement. As common in other brain disorders, the patient frequently attempts to explain away the embarrassing, incongruous behavior by composing a reasonable but fictitious scenario termed a "confabulation." Gelastic attacks are usually brief (a few seconds to a few minutes), but can occur up to 20 times per day. In some cases, laughing seizures are also associated with involuntary crying (dacrystic or quiritarian epilepsy), arm waving, knee slapping, and running (cursive epilepsy).

The case of "running epilepsy" is provocative evidence for a neurological link between emotional states (fear is most common) and removing oneself from harm's way.

Penetrating the thicket of research papers on this condition is challenging because gelastic epilepsy may be not one, but several disorders, linked by the common element of involuntary laughter. Note the variations in the reported symptoms. The characteristic paroxysmal laughter *may* or *may not* be associated with identifiable brain electrical (EEG) events, awareness of the laughter, or a seizure-produced emotional state. Furthermore, the gelastic attacks can occur before, during, or after generalized convulsive seizures, or in the absence of other seizures, and may originate in several brain regions, with the hypothalamus and temporal lobe of the cerebral cortex being implicated most often. One seizure type involves the involuntary production of laughter, awareness of the laugh-related events, but no emotional content—a type of seizure associated most often with tumors or lesions of the hypothalamus, thalamus, and brain stem. Seizures producing a jovial or other emotional state (aura) are sometimes associated with the temporal cortex. Gelastic seizures triggering both a jovial mood and laughter are exceedingly rare, because people are usually amnesic of the seizure related subjective state.

Electrical stimulation of the brain of a seizure-prone 16-year-old girl provides a rare and informative view of brain-driven laughter *and* a mirthful emotional state in a single person. The girl was being examined by Itzhak Fried and colleagues at UCLA to locate the source (focus) of her intractable seizures. (Laughter was not associated with these seizures.) Stimulation of a small area of her left frontal cortex (anterior supplementary motor area) reliably caused her to both laugh and perceive things as "funny." Each stimulation produced a different humorous scenario. When stimulated while looking at a picture of a horse, for example, she laughed and commented "the horse is funny"; when attending to people in her room, she observed "you guys are just so funny." Also intriguing was evidence linking smiling and laughter. Low levels of electrical stimulation caused her to smile, with higher levels triggering hearty guffaws.

The stimulated brain site probably triggered the emotional state common to both acts, which, in turn, generated the motor patterns of smiling and laughter. Previous cases of electrically driven laughter have involved the anterior cingulate and orbitofrontal cortex and the basal temporal lobe, with one study suggesting a neurological dissociation of the act of laughter (associated with the anterior cingulate) and mirthful experience (associated with the temporal lobe).

A strange and tragic seizure disorder involving laughter is sometimes found in children developing precociously. Imagine a big-for-her-age 3-year-old girl with pubic hair and the breasts and external genital development of an adolescent. Or a big 1-year-old baby boy who rumbles baby talk in a deep, resonant voice and has an enlarged penis subject to frequent erections. The boy delights in the company of older females (not men), whose skin he loves to stroke. By age 2½ years, he has heavy, dark pubic hair and the genitals of a typical 12- to 14-year-old. These are symptoms of precocious puberty caused by tumors (hamartomas) of the posterior hypothalamus, a condition associated with brief seizures of a petit mal type (with maintained consciousness), normal, then regressing intelligence, normal neurological status, and an early death, usually by the teen years. Laughter or crying sometimes precipitates and is maintained during seizures in several of the reported cases.

Reports from the Land of ALS is a remarkable personal account by Francis McGill of her life with amyotrophic lateral sclerosis (ALS or Lou Gehrig's disease), a progressive and fatal motoneuron disorder. As with famous astrophysicist Stephen Hawking and other victims of this disease, the intellect is spared as the nervous system gradually loses control of its muscles, leaving the patient able to sense and think clearly, but progressively unable to act on the world and control vital body functions. Some ALS patients experience a less well-known symptom, an instability ("lability") of emotional display that may include inappropriate and unfelt bouts of jocularity and stormy bawling outbursts that are as puzzling to the patient as to their family. During the precious two-year window during which she was still able to communicate, Francis McGill wrote of her descent into ALS, including the neglected theme of emotional turbulence. The following passages were excerpted from her chronicle:

> I was not as upset or as sad as my crying would imply, nor as uproariously amused as my uncontrollable laughter would indicate . . . Such episodes of laughter and tears may have slight connection with my actual frame of mind . . . In fact, I usually become very frustrated and angry at my inability to put a halt to such ridiculous behavior . . . I begin to smile, but the smile becomes an exaggerated grin, which attaches itself to my face and I have to use all of my powers of concentration to remove the embarrassing grimace. If I yield to the impulse, it becomes the onset to equally uncontrollable giggles, which in turn so embarrass me that I become angry, humiliated and subject to uncontrollable tears—it is a vicious, see-sawing circle.
>
> That which in another day and age, might have been a slight misting of the eyes or lump in the throat, is now harboring about as the screaming monster of a "bawl." . . . I am mortally afraid of squealing bawls. They destroy me—they weaken and crumble me . . . those deep debilitating agonizing episodes. They are not gentle rain. They are more hurricanes. For me, tears are no longer healing, but laughing is fun if it does not continue so long as to give me an aching solar plexus.

In regard to seeing her son:

> I often begin laughing at the sight of him . . . The sight of him pleasures me . . . And although I wish I had a more sensible and controllable reaction, since I can't seem to help it, I might as well enjoy it.

Francis McGill felt like a leaf buffeted by an inner storm, and the fight for emotional control left her tired and frustrated. Attempts to dampen her wild emotional swings by mentally focusing on something boring, like a chair, offered only partial relief. But she remained sane and strong in the face of her runaway emotional behavior, like a competent driver struggling to control a car that has no brakes and is perversely overresponsive to every tap on the accelerator or turn of the wheel. The mood swings she describes are rather like our own, except much stronger in intensity and duration, another case of the pathological being an extreme variant of the norm.

Her story dramatically illustrates the necessity of distinguishing between externally observable behavior and subjective mood states—her hyperactive emotional displays were not associated with a mood disorder as found in depression or mania. However, more suggestible ALS patients may be convinced by their excessive crying and family reaction that they actually are depressed, a self-fulfilling prophecy that makes an already-depressing circumstance more difficult to bear.

On another level, McGill and other patients with such symptoms present a novel challenge to the James-Lange theory of emotions that posits that mood is determined by the perception of the body's physiological state—their mood is only loosely associated with their intense emotional displays.

Francis McGill has a heavy burden in this section, speaking both for fellow ALS patients and those who have similar symptoms of emotional lability caused by other pathology. McGill is an articulate proxy for patients less lucid and able to render their subjective state. The exaggerated or inappropriate emotional display accompanied by appropriate emotional experience described here in ALS is characteristic of the pseudobulbar state, an occasional symptom of stroke, multiple sclerosis, and brain trauma. I will conclude with an anecdote that illustrates the power and social costs of emotional incontinence.

An ALS patient and her advocate from a local ALS support group were attending a court hearing to appeal a 50 percent reduction in the hours of an aide that assisted the patient in her home. While being questioned by the judge, the patient began giggling and laughing, a response that brought a quick gavel and the threat of a contempt-of-court charge. The laugher continued. When the advocate attempted to speak on her client's behalf, she too was threatened with contempt. Not deterred, the advocate bravely yelled back that the "contemptuous" laughter was a symptom of her client's disease. Reason prevailed, and the judge relented, but the patient lost her appeal, the first step toward being forced to leave her home and move into a nursing facility.

Clichés about madness are full of references to laughter: Mindless giggles, socially inappropriate cackles, or guffaws in response to a private script all trigger powerful social sanctions. Even witnesses who don't know the rules of laughter vigorously respond to their violation. These matters are the subject of one of history's notable medical consultations—that between the founder of atomic theory and the father of medicine.

Democritus of Abdera (460–370 B.C.) was one of the most original and powerful intellects of antiquity. He is best remembered for the undeniably great idea that the cosmos is composed of particles called atoms. Although generally cheerful, Democritus was also known to be sarcastic, crisp, and unconventional, and did not suffer fools gladly. His fellow Abderites became concerned when the great sage began acting queerly, laughing at everything. Fearing that he might be going insane, they sought the assistance of Hippocrates, a renowned medical specialist. After a lengthy, ironic conversation with Democritus about the foibles and absurdities of human nature, Hippocrates assured the townspeople that their sage was quite sane—perhaps too sane for comfort. Democritus became known as the "laughing philosopher," in contrast to his somber Ionian neighbor, Heraclitus ("We never step into the same river twice"), the "weeping philosopher."

Democritus was lucky. Until recent times, inappropriate laughter accompanied by other odd behavior could get you run out of town, ostracized, or burned at the stake for witchcraft. It can still get you locked up, drugged, or made a social pariah. The association of laughter with a person's social and emotional integrity is not arbitrary—laughter signals activity of the brain regions most implicated in mental disorders. The presence of uninhibited, inappropriate laughter in mania and hysteria is well known and discussed elsewhere (Chapter 7).

Unusual laughter is also a symptom of certain subtypes of schizophrenia. In Emil Kraepelin's first descriptions of dementia praecox ("early dementia"), as schizophrenia was once known, he comments on its victims' "silly laugh," noting the absence of corresponding joyous humor. The laughter "is unrestrained, appears on all occasions without the least provocation and is altogether without emotional significance." Eugen Bleuler follows up, commenting that the "compulsive laughter" of schizophrenia is a "soulless mimic utterance, behind which no feeling is noticeable." He coined the term "parathymia" to describe the inappropriate affect in schizophrenia, characterized by abnormal laughter or sorrow. "A particularly frequent form of parathymia is represented by unprovoked or inappropriate laughter." "Usually the patients themselves do not know why they are being compelled to laugh." According to the classification scheme of the latest *Diagnostic and Statistical Manual of Mental Disorders* (DSM-IV), such laugh symptoms are most typical of schizophrenia of the disorganized type. When I worked with such patients as a neuropsychology trainee at a VA hospital in the mid-1960s, they were most often classified as "hebephrenic" schizophrenics.

Although I became personally acquainted with many patients on the wards during interviews and testing, in my small jazz band (music therapy), and psychodrama sessions, I never got to know the hebephrenics, and had difficulty remembering them as individuals. It's difficult to penetrate a veil of silliness and indifference to become acquainted with a person, whether the state is produced by alcohol, drugs, or, in the case of schizophrenia, an aberrant neurotransmitter bath arising within the brain. Lacking a coherent sense of self and contact with the environment, these unfortunate patients may not reach the social threshold for personhood. In this train wreck of a mental illness, little may be left of the person to know. Borrowing from T. S. Eliot, that is the way the world ends, not with a bang but a giggle.

The prefrontal lobotomy was a tragic misstep in the history of psychiatry, born of the frustration and futility of treating schizophrenia in an age before the development of effective antischizophrenia (neuroleptic) drugs. During the 1940s and 1950s, thousands of seriously disturbed and unruly patients in mental hospitals were subjected to this crude and untested surgical procedure, which, in varying degrees, damaged the frontal cortex and its connections to the rest of the brain, resulting in major changes in intellectual, emotional, and social behavior. After surgery, some lobotomized patients experienced bursts of apparently involuntary laughter and crying.

The "laughing spells" of four lobotomized schizophrenic patients are reported by H. C. Kramer in a paper published in 1954 during the heyday of the procedure. The first was an unfortunate woman diagnosed as "catatonic" who had received 69 shock treatments prior to her surgery. Like most candidates for lobotomy, she was troublesome to staff and other patients. After surgery her behavior changed, but not necessarily for the better, for she became "overactive, overtalkative, elated, and exuberant to the point of annoyance." She also developed bouts of laughter that she was unable to control. Her inappropriate merriment interfered with life on the wards, provoked complaints from other patients, prevented work, and sometimes prompted her removal from social settings. The other three cases (one "catatonic" and two "hebephrenic" schizophrenics) experienced similar postoperative bouts of uncontrollable laughter and associated problems.

The most unusual report of postoperative euphoria concerns an "auto-lobotomy," a case of do-it-yourself neurosurgery by gunshot wound, the result of a botched suicide attempt. The radical procedure was said to have reduced the patient's depression, but at the cost of lost initiative, increased facetiousness, inappropriate laughter and crying, and incontinence.

The outcomes of these surgical cases are generally consistent with what his known about accidental damage to the frontal cortex, an event that can produce inappropriate jocularity (*Witzelsucht*), tasteless humor, silly and euphoric behavior (*moria*), wild swings of exaggerated emotional expression (laughing and crying), and sometimes dulled emotional experience. Because frontal lobe damage may not produce the obvious disorders of speech, movement, and sensory function associated with lesions of other cortical lobes, early investigators erroneously concluded that the frontal lobes played some relatively nonspecific, partially expendable role. In fact, the frontal lobes function more in the subtle realm of thought and ideas (sometimes pathological) than of sensation and action. But the effects of frontal-lobe trauma, although sometimes subtle, can be devastating. Life can't proceed in an orderly manner without the critical executive functions (timing, planning, anticipating consequences, emotional control, etc.) associated with the frontal area.

A challenge in the rehabilitation of some frontal lobe patients is teaching them to control their emotional outbursts. Imagine the frustration of trying to meet the cognitive demands of a job, yet being held back by uncontrollable laughter. This is not as trivial an issue as it may seem on first hearing. Recall the courtroom anecdote involving the ALS patient. Laughter is a potent signal that triggers a strong emotional reaction in those around us, especially if it's inappropriate.

The medical literature lists many conditions that disrupt the control and context of laughter, but lacks a single report of a certain dramatic, irreversible pathology of laugh production that strikes tens of thousands of victims yearly. I learned of it quite by chance when a wheelchair-bound student offered to provide a sample of laughter for my study. The student was quadriplegic as a result of an automobile accident that damaged his spinal cord in the low cervical region, depriving him of sensation and voluntary movement of his legs, part of his arms, and trunk below the chest. The best laugh he could manage was a rather feeble, sustained expiration or "haaaaa"—he lacked the explosive chain of "ha-ha-ha"s that are typically blasted out during guffaws. He could, however, voluntarily speak a weak, unlaugh-like "ha-ha-ha." The student also had a more serious deficit—he couldn't manage a vigorous and "productive" cough, an explosive airway maneuver with some laugh-like properties that is essential for clearing the lungs. Interviews with other quadriplegics confirmed these findings. Breathing and speaking (but not laughing and coughing) were spared in these individuals, with persons having higher-level cervical lesions experiencing the greatest deficits. Spinal lesions above cervical level 3 paralyze the muscles of respiration, speech, and laughter, and are fatal unless the person is artificially respired.

The loss of laughter but not speech in the "natural experiment" of quadriplegia is definitive evidence that normal laughter is not simply a matter of speaking "ha-ha-ha." (Evidence based on the differential conscious control of laughter and speech was reviewed in Chapter 3.) Laughter and speech use different mechanisms that are differentially affected by spinal lesions. The sparing of speech in quadriplegia contrasts with the genetic disorders of Angelman and Rett, in which speech is usually absent but laughter, coughing, and other respiratory maneuvers remain normal.

A suite of six neurological conditions is now considered, five of which have laughter as a symptom (Rett and Williams disorders, Wilson, Alzheimer, and Pick diseases) and one that is notable because it does not (Tourette disorder). Although these conditions, with the exception of Alzheimer disease, are rare, their inclusion rounds out our collection and may prompt interest in laughter as a symptom of pathology.

Imagine the concern of a parent who hears the uproarious, midnight laughter of a just-wakened daughter sitting alone in her bed in a darkened bedroom. One mother asked such a laughing child, "Have you been playing with angels?" More anxious parents found the solitary night laughter "scary" and "unnerving," noting its "demonic" quality. This strange solitary laughter might suggest spiritual possession in an earlier era. But such laughter is symptomatic of 83 percent of young girls (no boys are affected) with Rett disorder. (The already discussed Angelman disorder also features laughter and sleep disturbances, but not night laughter.) Rett girls show normal or nearly normal early development until 6 to 18 months of age, followed by a regression of communication skills, loss of purposeful hand use, and the emergence of stereotyped hand movements. Other symptoms may include seizures and disorganized breathing (hyperventilation, breath holding). Tragically, Rett parents can only look on helplessly as this disorder steals the early promise of their young daughter's life. Rett girls experience a full range of emotions and exhibit engaging personalities, important in maintaining the unusually strong commitment of parents and caregivers. Parents often speak of the "spiritual depth," "purity," "innocence," and "inner joy" of their Rett daughters, a critical counter to their often chronic crying and irritability.

The ready laughter, cheery exuberance, bright eyes, and broad smiles of Williams disorder children also captivate parents, teachers, and acquaintances. But in Williams disorder, the prominent laughter is an amplification of the normal, the expected consequence of a high activity level and upbeat, gregarious nature. Williams disorder is a rare, sporadically occurring neurobehavioral disorder associated with a deletion on chromosome number 7 that affects males and females equally and does not run in families. Williams disorder offers an informative contrast with the better-known affliction of autism (Chapters 3, 6). While autistic persons are aloof and lack social insight, Williams people err in the other direction—they are excessively social. Empathetic Williams children may be easily moved to contagious tears or rowdy laughter by the actions of their classmates, and they may indiscriminately seek conversation and even hugs from strangers as well as friends. Sadly, Williams people have trouble forming friendships—difficulties may be associated with their unnuanced and indiscriminate social advances, unusual, elfin appearance, impulsiveness, and learning problems. Along with their high degree of social and emotional sensitivity comes a remarkable mix of cognitive strengths and weaknesses, with the deficiencies preventing them from leading an independent life. An impressive vocabulary may be coupled with the inability to write, to count objects on a page, or to tie one's shoelaces. In one case, a vocal repertoire of hundreds of memorized songs in different languages was linked with an inability to subtract or judge the distance of an oncoming automobile.

Wilson disease (hepatolenticular degeneration) is a rare progressive disorder of copper metabolism associated with an autosomal recessive gene. In contrast to the exogenous origin of manganese poisoning (*masque manganique*), this case of metal toxicity has a genetic basis. The accumulation of copper in the brain (especially the basal ganglia) leads to tremor, muscular rigidity, involuntary choreiform (dancelike) movements, dementia, and of primary interest here, emotional lability, including involuntary laughing and crying. In the liver, the high copper level causes cirrhosis. Although Wilson disease eventually produces dementia, it's considered a movement disorder because its first symptoms are jerky movements. Symptom onset is usually between 10 and 25 years of age. In contrast to the grim prognosis of manganese poisoning and most progressive neurological disorders, if treated early, the onset or progression of Wilson disease can be blocked by a drug (D-penicillamine) that binds and permits the excretion of copper.

Alzheimer disease is a degenerative brain disorder that first appears as a memory loss but progresses into a generalized dementia that slowly robs victims of their personhood and social bonds with family and friends. The families of Alzheimer victims often witness the less publicized emotional symptoms of Alzheimer disease, including pathological laughing and crying. A recent study of 103 Alzheimer patients found that 39 percent had pathological affect, with 25 percent showing crying episodes and 14 percent showing laughing or mixed laughing and crying episodes. Although emotional outbursts were sometimes associated with an underlying mood disorder (depression), about half of all Alzheimer patients with abnormal affect showed no relationship between mood and behavior, perhaps because they do not monitor their emotional state and displays—they are not in touch with their feelings.

Pick disease is a progressive cortical dementia of unknown cause with many Alzheimerlike behavioral qualities, but with social instead of memory-related anomalies as its first symptoms. The most obvious early signs of Pick disease include abnormal euphoria, a deterioration of interpersonal skills, and personality changes. The memory, cognitive, and linguistic deficits typical of early Alzheimer disease appear only later. Pick and Alzheimer diseases can also be distinguished at the gross and microscopic levels, with Pick exhibiting more focal cortical atrophy, especially in the frontal and temporal regions.

Laughter is absent from lists of Tourette disorder symptoms, a fact notable because its symptoms often include the somewhat similar utterances of coughing, an explosive expiration, and cursing, another emotionally loaded vocalization. (Normal laughter is present.) A single explosive laugh-note ("ha"), however, may not be recognized as laughter. Tourette disorder is a subcortical movement disorder featuring vocal and motor tics, most often of the face and head, that the patient is compelled to perform. The tics can be inhibited voluntarily for short periods, after which they rebound to higher than normal levels. In more severe forms of Tourette, movements may be complex and vigorous, involving touching, hitting, and jumping. Other symptoms may include the mechanical repetition of observed acts (echopraxia) or sounds (echolalia), and its signature symptom, bursts of obscene speech (coprolalia). Given that echolalia is sometimes a symptom of severe Tourette disorder, it would be informative to know if this disposition extends to contagious laughter and yawning. Indeed, might echolalia in Tourette and elsewhere involve pathology of the process responsible for the normal contagiousness of laughter and yawning?

In "The President's Speech," a clinical tale by neurologist Oliver Sacks, a group of aphasics was observed roaring with laughter while watching a televised speech of then President Ronald Reagan. It was extraordinary that this group was being emotionally moved by the speech, for each of them had a severe global aphasia that made the words of speech unintelligible and speech impossible. At other times, these profoundly impaired people sometimes grasped enough of the gist of speech to impress friends, family, and hospital staff with their apparent comprehension. In fact, Sacks's aphasics were responding to the *tone,* not the meaning of the words, or grammar of speech. After all, you need not parse and analyze a sentence to know if a speaker bears good or bad news, or is happy, sad, or angry.

The tonal, or prosodic, character of speech carries much of its emotional context, helped along by facial gestures and body language. We are especially aware of these tonal cues when they are absent, as in machine-produced computer-speak. Global aphasics of the type considered by Sacks usually have large left hemisphere lesions and can neither produce nor comprehend speech. (The reciprocal disorder [tonal agnosia], the inability to perceive tonal cues, but with an intact ability to produce and comprehend words, is associated with lesions of the right temporal cortex.)

Oliver Sacks's clinical tale describes how global aphasics use the rich nonlinguistic cues in speech to detect and laugh at the tonal and gestural incongruities in the president's speech. But his anecdote has other intriguing laugh-related features, not the least of which is that these presumably speechless patients are laughing. This, in itself, is not surprising, because otherwise speechless aphasics are known to laugh, cry, produce the emotion-laden burst of an expletive, and sometimes even sing. And much normal, everyday laughter is not a response to humor, but to nonlinguistic prosodic and social cues of the sort used by Sacks's aphasics. This vocal sparing in aphasia is also evidence of the already noted segregation, but not complete independence of the laughter and speech mechanisms.

The laughter of aphasics is likely to yield other insights. For example, some aphasics may "punctuate" the speaker's speech stream with their own laughter, just as nonaphasics do, and even laugh appropriately after the punch lines of jokes. The "timing" of a comic routine involves setting up and maximizing these paralinguistic signals and gestures, cuing the punch line, and giving space for the audience to laugh.

In a series of innovative studies, Hiram Brownell, Howard Gardner, and their colleagues probed the effects of unilateral stroke damage on humor comprehension. Although these pioneers don't lead us all the way to the promised land, they do blaze a trail into the neurological and psychological wilderness. Along the way, they discovered that unilateral lesions to the left and right hemispheres produce very different senses of humor.

Their first study examined humor comprehension and appreciation in patients with lesions of the left or right cortical hemisphere. Patients were asked to choose the funniest of four cartoons, either with or without captions. The left-hemisphere lesion patients best understood cartoons without captions, a result consistent with the well-established language specialization of the left hemisphere. In contrast, the right-hemisphere patients (with intact left-side language areas) did best on cartoons with captions. But other aspects of patients' behavior were more telling. The left-side group behaved relatively normally during the task, whereas the right-side group was either unresponsive (even when they understood the joke), or laughed indiscriminately at every item. Although humor comprehension was impaired in both groups, the right-side lesions were generally more disruptive.

Another study used a joke-completion procedure to examine humor comprehension in patients with right-hemisphere lesions. The patients were read the beginning of a joke and then asked to select the funny ending from choices that were read and presented on a card. This procedure builds upon two components presumed to be present in jokes: surprise and coherence. As suggested by the incongruity theorists of Chapter 2, a joke begins by establishing an expectancy that is violated by the surprise of the punch line. However, to be funny, the punch line must be coherent with the joke's storyline, otherwise the punch line would be an unfunny non sequitur.

Review the following joke.

"The neighborhood borrower approached Mr. Smith on Sunday afternoon and inquired, 'Say, Smith, are you using your lawnmower this afternoon?'

'Yes, I am,' Smith replied warily.

Then the neighborhood borrower replied, '______.'"

The subject completes the joke by selecting one of the following:

Correct punch line (surprising and coherent):

"Fine, then you won't be wanting your golf clubs. I'll just borrow them."

Straightforward ending (coherent but no surprise):

"Do you think I could use it [the lawnmower] when you're done?"

Non sequitur (surprising but not coherent):

"You know the grass is always greener on the other side."

As expected, the right-side lesion group made more errors than the neurologically intact control subjects. Of more interest was the pattern of errors made by the two groups. The right-side lesion group was uniquely deficient in the coherence dimension of jokes—they were incapable of reinterpreting the narrative using information provided by the punchline to make sense of the whole joke. However, they understood the significance of surprise, showing an abnormal predilection for non-sequitur endings of the "grass is greener" variety. The "grass is greener" option may raise eyebrows, but will generate few laughs.

Prathiba Shammi and Donald Stuss built upon this pioneering work by identifying the frontal lobe as the right hemisphere site critical for humor appreciation. In addition to taking Brownell's Joke and Story Completion Test, their subjects rated the funniness of humorous and neutral statements (Humorous example: Sign in a Tokyo hotel—GUESTS ARE INVITED TO TAKE ADVANTAGE OF THE CHAMBERMAID.) and tried to pick funny cartoons from a series of neutral illustrations. Patients with selective right frontal-lobe damage performed poorly on both verbal (jokes) and nonverbal (cartoons) tests of humor and reacted with little laughing and smiling. But humor detection was not a focal defect, reflecting the complexity of humor processing. Diminished humor appreciation was associated with impaired working memory, mental flexibility, abstract ability, visual search, and emotional expression. Although obviously important to humor, the right frontal lobe is not the "humor center," but the neurological confluence of the numerous cognitive and affective inputs necessary for humor and associated laughing and smiling.

The ability to understand punch lines and "get" verbal and nonverbal humor tap high-level cognitive skills involved in emotional intelligence, sense of self, and ability to write the scripts that guide our lives. As such, humor provides a nontraditional but useful measure of social and cognitive competence. This line of research also shows how cognitive neuroscience can approach difficult philosophical and psychological problems concerning the nature of jokes (e.g., the roles of coherence, incongruity, and surprise), while revealing how the brain processes information to create humor.

People are said to die from a broken heart, but can death result from laughter or joy? Aside from the tickle torture described in Chapter 6, how risky is laughter? Galen (A.D. 129–199) considered joy to be more dangerous than anger, and history provides examples of people falling dead at moments of great joy or triumph. A sample of such fatal moments is provided in the 1896 text *Anomalies and Curiosities of Medicine* by George Gould and Walter Pyle, a treasure trove of the grotesque, titillating, and simply odd in medicine. A woman fell dead at the sight of her son she thought dead. An overjoyed Frenchman died upon release from prison. A niece of Leibnitz died when learning of a large inheritance of gold from her deceased uncle.

Exuberant laughter has also claimed a few victims. The nineteenth-century English novelist Anthony Trollope (1815–1882) had a stroke while laughing in response to readings from Anstey's comic novel *Vice Versa.* "For a while, 'Uncle Tony roared as usual,' and then suddenly Tilley and Edith realized that as they were laughing he was silent." Trollope was left speechless and paralyzed on his right side, and died several weeks later.

Laughter has also claimed victims in more extraordinary circumstances. An American frontiersman came home to find his wife and children dead, scalped and mutilated by Indians. He laughed uncontrollably, exclaiming, "It is the funniest thing I have ever heard of," until he died of a ruptured blood vessel. This anecdote appeared in an 1897 paper co-authored by G. Stanley Hall, one of the founders of American psychology.

A woman of 58 was hospitalized and suffering from pseudobulbar palsy, a condition associated with exaggerated emotional displays. One morning she experienced a severe headache, her eyes rolled upward, and her body became racked by intense, uncontrollable laughter during which breathing was nearly impossible. A morphine injection had no effect. The terrible laughter continued unabated for one and a half hours after which she collapsed, entered a coma, and died 24 hours later from a massive cerebral hemorrhage. This is one of the most straightforward clinical reports of someone "laughing to death." But the lethal role of laughter is seldom unambiguous. For example, a young man of 25 suffered an embarrassing attack of uncontrollable laughter at graveside during his mother's funeral. He later sought medical attention to deal with recurring bouts of such laughter. While hospitalized for tests, he was found dead with a ruptured aneurysm. Was his laughter a consequence of aneurysm formation, enlargement, or rupture? Or did his laughter rupture or play a direct role in the formation of the aneurysm? Or were the laughter and aneurysm causally related to some third variable?

Although laughter may be the best medicine, it's not totally free of undesirable side effects. Consider the 62-year-old Massachusetts man who laughed so hard during the *Seinfeld* television show that he fainted at least three times. Once he went blind temporarily. On another occasion, "he fell face first into his evening meal and was rescued by his wife." Fitter souls need not fear that a comic onslaught will leave them facedown in their linguine. The man in question was predisposed to cardiac problems—he had high blood pressure and cholesterol levels, smoked, and already had a coronary bypass. A clinical checkup revealed that several arteries supplying blood to his brain were clogged. When he laughed hysterically, he spontaneously performed a so-called Valsalva maneuver—a bearing-down movement (as in a bowel movement or when lifting a heavy load)—that compressed his heart and further compromised the already inadequate blood flow to his brain. Balloon angioplasty opened his clogged arteries, and an implanted stent kept them open. He is now laughing without a relapse of "Seinfeld Syncope." (Syncope is a sudden loss of consciousness due to a decrease of blood supply to the brain.)

As with Seinfeld Syncope, these cases of lethal laughter probably involve people already predisposed to the ill effects of any kind of exertion, including those with heart disease, hypertension, and aneurysms. Such laugh victims are notable because of their novelty. Although comparative statistics are lacking, rage has probably felled many more victims than fits of laughter. Laughter remains highly recommended for a long and good life.

The problem with most existing clinical studies of laughter is that the majority of them aren't specifically about laughter. Laughter is often an afterthought of clinicians in pursuit of other more familiar problems. And when laughter is mentioned, there may be only a few intriguing asides that are lost in a sea of competing data. I offer a two-step solution. First, laughter must be recognized as a significant topic for clinical study, a position consistent with its prominence in daily life. Second, studies of laughter must be conducted in a more systematic manner, examining an agreed-upon set of variables, using a common metric, method, and terminology. This book is a source of such variables and norms derived from observations of thousands of normal subjects. By describing what is normal about normal laughter, we can describe, for the first time, what is abnormal about abnormal laughter, including that of people with neuropathology, psychopathology, and sensory impairment (blindness, deafness).

The following Laughter Inventory provides a list of characteristics that can be used to describe the laughter of normal and abnormal subjects. Consult the cited chapters for details about specific inventory items. Most items are based on observations of ongoing behavior and do not require formal testing. Taken together, the inventory items summarize laughter as a vocalization, perceptual stimulus, linguistic and social event, and response to tactile stimulation.

LAUGHTER INVENTORY

- *Laugh structure:* a measure of vocalization as described by the laugh-notes per episode, duration of laugh-notes and internote-intervals, and the frequency, amplitude, and harmonic composition of laugh-notes (Chapter 4). (This is the only measure that requires diagnostic equipment.)

- *Linguistic context of laughter:* an index of the integrity of the punctuation effect in both speaker and audience, the tendency of laughter to be placed at the end of phrases and other pauses in speech (Chapter 3).

- *Social context of laughter:* the proportion of laughter occurring in social as opposed to solitary settings, and the appropriateness of laughter's social circumstance (Chapter 3).

- *Humorous context of laughter:* the style of comedic stimuli that triggers laughter or is used to produce laughter in others (Chapter 3).

- *Contagiousness of laughter:* an index of the responsiveness to laughter produced during conversation or to that produced by a device such as a laugh box (Chapter 7).

- *Tickle-evoked laughter:* the amount and quality of laughter produced when being tickled or when tickling another person (Chapter 6). (This measure is most appropriate for young children, and the tickling should occur between individuals whose relationship includes consensual touching.)

We now come full circle and entertain the tough questions about abnormal laughter posed in the chapter's introductory quotations. Why indeed was Titus Andronicus laughing in a time of tragedy? And what gave the laughter heard by Jane Eyre its demoniac quality? Both are perfectly legitimate questions, but ones that challenge the power of the present analysis.

The spookiness of the distant laughter troubling Jane Eyre was based solely on acoustic clues. She responded to a nuance of the skeletal elements that give the "laughness" to laughter. Our ability to hear laughter as mirthless, tragic, eerie, sly, merry, sardonic, raucous, etc., is testament to the exquisite sensitivity of our auditory system to vocal shading, a standard that humbles the cruder hardware of acoustic analysis. Indeed, the disembodied laughter in the novel would probably be judged "normal" by the rough standards of laugh structure on the Laughter Inventory. (Anomalies in social, linguistic, and other dimensions would be detected when more becomes known about the laughter.) In the pursuit of laughter's universal sonic signature, the tactical decision was made to delay dealing with the daunting *Jane Eyre* problem, the bewildering variations of the basic underlying pattern. The difficulty of defining such tonal shadings is not unique to laughter, a fact well known to scientists seeking the acoustic roots of irony, sarcasm, and other subtle messages in speech.

What, then, can be said about the tragic Titus Andronicus laughing inappropriately at the receipt of his sons' heads and his severed hand in a box? Like the *Jane Eyre* problem, it depends on the contrast with "normal" laughter. Unlike the eerie laughter from *Jane Eyre,* Andronicus's laughter may be normal in form, but is deviant in the grotesque stimuli that triggered it. (This is an anomaly of the social and humorous contexts in the Laughter Inventory.) An attempt to account for this inappropriate laughter by appealing to hypothetical mechanisms (e.g., "emotional overflow"), or explanation by naming (e.g., "nervous laughter"), would bring us dangerously close to psychobabble or its more pretentious biological cousin, neurobabble. Andronicus does little better in his self-analysis—"Why, I have not another tear to shed." A more rigorous and poetic account of Andronicus's behavior is suggested by the intimate neurological relation between laughter (his paradoxical action) and crying (the more appropriate response). The clinical record provides several cases of brain damage and disease that produce inappropriate laughter and crying, with cases of emotional lability as in ALS producing wild, undamped swings between the two. Although laughter and crying are considered polar opposites of the emotional spectrum, they are neurologically linked and share the features of tearing and rhythmic vocalization. Our woeful Roman general of the Shakespeare play reminds us of prescient artifacts of a more ancient stage. The neurobehavioral relation between laughing and crying has been immortalized as the symbol for drama, the twin masks of Greek comedy and tragedy.

NINE

Laughing Your Way to Health

A merry heart doeth good like a medicine: but a broken spirit drieth the bones.

—Proverbs 17:22

And frame your mind to mirth and merriment, which bars a thousand harms and lengthens life.

—Shakespeare, *The Taming of the Shrew*

uthorities from the Bible to William Shakespeare and *Reader's Digest* remind us that "laughter is the best medicine." Laughter seems an ideal treatment that makes us feel good, is free, has no bad side effects, and is even contagious, bursting forth in a noisy and benevolent chain reaction. Print and broadcast reporters encourage our enthusiasm for medicinal laughter in upbeat and often frothy stories with titles like "Laughing Your Way to Health" or "A Laugh a Day," promotions of a tentative hypothesis that many people hope is true. Going unsaid in such reports is a basic and perhaps jarring truth: *Laughter no more evolved to make us feel good or improve our health than walking evolved to promote cardiovascular fitness.*

The idea that laughter is a calisthenic for body and soul has become so pervasive that we tend to overlook the fact that laughter evolved because of its effect on others, not to improve our mood or health. The presumed health benefits of laughter, like those of walking, are secondary, coincidental consequences of an act evolved in the service of another function. And what are we to make of the effects of humor? Humor and joking are modern cognitive and linguistic triggers of laughter, products of recently evolved cortical bells and whistles superimposed upon laughter's ancient neurological core. Does a ready laugh, cheery disposition, and good sense of humor confer a longer or better life? And if humor proves beneficial, who benefits most, its dispenser or consumer?

At the risk of becoming a curmudgeon and New Age pariah, I suggest that we begin by taking a scientific perspective toward these issues, assuming that laughter has no therapeutic value at all (the "null hypothesis"), until we learn otherwise. A broad and open mind is a necessary antidote to the unabashed advocacy and entrepreneurism infecting so much contemporary reporting of laughter's wondrous medicinal effects. I have witnessed this media bias first hand. Over the years, I have been contacted by many print and broadcast reporters working on yet another tired story about "laughing your way to health." My message that the literature about laughter and health is not all that it seems, and that laughter did not evolve to make us healthy (although it may indeed contribute to good health), was as welcome as a skunk at a picnic, and certainly not what these purveyors of popular culture thought their audience wanted to hear. With the media contributing to the confusion, it's difficult to cut through the hype and get straight talk about laughter, humor, and health. The following account attempts to separate fact from fancy and provide an unbiased guide to this fascinating but sometimes confusing topic.

The idea that laughter is therapeutic has existed since antiquity but in recent decades has been popularized by Norman Cousins in his

1976 article "Anatomy of an Illness (As Perceived by the Patient)," published in the *New England Journal of Medicine*, and expanded into a book in 1979. In this work, Cousins describes his affliction with a painful and life-threatening degenerative disease (ankylosing spondylitis), and his successful self-treatment with vitamin C, the Marx Brothers, and episodes from the old and very funny television series *Candid Camera*.

In collaboration with his physician, Cousins removed himself from the not-so-tender mercies of the hospital and checked into a hotel, where he improvised a therapeutic regime for body and soul, including a healthy dose of humor. He claimed that ten minutes of belly-laughter provided him at least two precious hours of pain-free sleep and other desirable health benefits. As editor of *Saturday Review* and a well-connected man of letters, Cousins knew how to tell a good story and maximize its impact. Cousins's book became a best seller, and he parlayed his newfound status in the medical community into a faculty position at the UCLA School of Medicine and became a widely sought speaker on medical matters from the patient's perspective. Late in life he moderated his laugh-your-way-to-health message, noting that humor should be considered a metaphor for the entire range of positive emotions. Although aspects of Cousins's inspiring recovery have been challenged (he may have experienced spontaneous remission, may not have had ankylosing spondylitis, vitamin C may not have had the presumed effect, etc.), his eloquent reminder that patient and physician attitudes, expectations, and behavior contribute to healing has lasting value and has become a cornerstone of holistic medicine.

Cousins's timing was ideal. The medical and nonmedical world had been prepared for his message of a mind-body linkage by the wave of biofeedback research during the 1960s and by even earlier seminal work by scientists Hans Selye (*The Stress of Life*, 1956) and Walter B. Cannon (*The Wisdom of the Body*, 1932). By the mid-1970s, it took no leap of faith to believe that love, hope, and laughter could produce a benevolent body chemistry conducive to health. The corrosive effects of stress, anxiety, and anger were already being catalogued enthusiastically by the medical community. By the 1990s, we have ample evidence of the public acceptance of the power of positive emotions. Witness the Robin Williams hit film about clown/physician Patch Adams and the best-selling books by surgeon Bernie Siegel (*Love, Medicine and Miracles*) and self-help guru Deepak Chopra (*Quantum Healing*). But what does current research actually say about the benefits of laughter and humor?

Case studies and anecdotal evidence in the style of Cousins offer promising leads and are useful starting points for biomedical research. The hard part is what comes next—shaping vague ideas into testable hypotheses and developing research methods and designs for a new area having no established priorities and traditions. The step from essay writing to the grunt work of empirical research is a large one—and one that has been taken by few scientists. Research on the health effects of laughter is conducted by a few dedicated individuals with heterogeneous interests and backgrounds, who are joined occasionally by specialists in relevant disciplines (i.e., behavioral medicine, psychophysiology, psychoneuroimmunology, social psychology) who have a passing interest in laughter. Given the prominence of laughter in our lives, it is surprising to learn that there is no large research laboratory dedicated to the study of laughter and no federal research initiatives to support one—politicians would surely ridicule any such proposal. This low level of support extends to all positive emotions and greatly influences the quantity, quality, and continuity of research.

The exploration of medicinal mirth begins with the description of laughter's physiological profile, a work still in progress. Laughter is the kind of powerful, bodywide act that really shakes up our physiology, a fact that has motivated speculations about its medicinal and exercise benefits since antiquity. During vigorous laughter, we take a deep breath, throw back our head, stretch muscles of the face, jaw, throat, diaphragm, chest, abdomen, neck, back, and sometimes the limbs, and exhale in explosive, chopped "ha-ha-ha"s. When our breath is exhausted, we often take another deep breath and start the cycle all over again. If sustained and vigorous, laughter may cause our sides to ache, our eyes to tear, and the occasional celebrant to tumble to the floor.

Pioneer laugh researcher William Fry made a rough estimate of the work of laughter using himself as a subject and heart rate as a measure of exertion. Fry found that it took 10 minutes of rowing on his home exercise machine to reach the heart rate produced by one minute of hearty laughter. He also noted that some exuberant laughers produce heart rates over 120 beats per minute for intervals over three minutes. Despite the obvious work of hearty laughter—we can laugh until our sides ache, and our heart rate and blood pressure surge—laughter seems not to be associated with significant metabolic need or cardiovascular risk, and we quickly return to a relaxed state. (But consider the rare cases of fatal laughter in Chapters 6 and 8.) Fry observed that the sustained expiration of laughter depletes no more blood oxygen than ordinary conversation. Although its caloric cost and other details are yet to be specified, laughter may provide a gentle form of aerobic exercise. But for laughter to be a serious form of exercise, it must be sustained beyond the occasional yuks triggered by the likes of Laurel and Hardy. The only stimulus I know that will reliably produce sustained laughter is tickle, and we learned of its tortuous downside in Chapter 6. Will any laughter produce an exercise effect, or must it be triggered by comedy or perhaps contagion? These are the kind of questions that go unanswered in most studies of laughter effects.

Madan Kataria of Bombay, India, is taking the exercise proposition very seriously. He has popularized an ancient yoga breathing exercise based on laughter, transforming it into a booming enterprise, Laughing Clubs International. Within one year of founding the first small group in 1995 that met in his home, the club has grown explosively to over 100 branches which meet daily in public parks—28 clubs are in Bombay alone. In Kataria's first group, members exchanged jokes, the quality and quantity of which quickly declined. Then he discovered that you can dispense with the jokes altogether—all that was necessary was to laugh, and soon everyone would join in the chorus of contagious laughter.

After starting with a warm-up of unison, "ho-ho, ha-ha,"s the group moves on to more esoteric variations with mouth open and closed. A favored laugh posture is standing with arms raised above the head, also a posture of Christian Pentecostals during worship and performance of their own contagious Holy Laughter. Instead of communing with God, Kataria's followers strive for the more secular objectives of lowered blood pressure, enhanced respiratory function, and general fitness. The Laugh Clubbers are certainly getting entertainment and companionship during their workouts, but until we know more, we must reserve judgment about the remainder of their physiological wish list.

Many research volunteers have watched films in the name of laugh research, experiencing the joys of a rollicking comedy or the boring ordeal of a dreary control flick about the beauty of the Scottish coastline. Hardy subjects also have urinated, spit, or bled to provide fluids for chemical analysis. Others provided less personal data about their blood pressure, heart rate, skin conductance, temperature, and respiration. Although submitting to these procedures may seem a steep price of admission for an uncertain cinematic experience, the stakes are the discovery of laughter's elusive physiological profile.

Lennart Levi of the Karolinska Institute in Stockholm published the first such biological movie review of a comedy in 1965. He recruited 20 female office clerks who watched several films, before, during, and after which they dutifully provided urine samples for analysis. The intensity of the subjects' emotional arousal was measured in terms of urine epinephrine (adrenaline) and norepinephrine (noradrenaline), hormones that increase heart rate, blood pressure, and metabolic activity. Epineprhine and norepinephrine are members of the catecholamine family of hormones and neurotransmitters of the sympathetic nervous system, the body's "fight or flight" system. When the catecholamines start flowing, the body is "stressed" and getting ready for emergency action. The results were surprising—both comedy *and* intense drama produced physiological arousal. Epinephrine and norepinephrine levels were elevated during the viewing of the comedy *Charley's Aunt*, the stressful war film *Paths of Glory*, and the gruesome ghost story *The Devil's Mask*, but not during a bland natural scenery film.

Other investigators followed Levi's lead and published their own biological movie reviews, but many studies of this genre are dated, report different measures, have methodological inadequacies, and reach no consensus. No amount of scholarship can piece together a coherent physiological picture of laughter or humor from this hodgepodge of odd parts. Investigators often disagree about the most basic results. Does laughter or humor increase or decrease physiological arousal or have no effect at all? Also, no study separates the effects of laughter from those of humor. (Laughter without humor can be produced by contagion, tickle, or mirthless, forced laughter. Humor without laughter can be produced by inhibiting laughter during a comedy performance.) No study controls for the possibility that presumed effects of laughter or humor may really derive from the playful social settings associated with these behaviors, not the acts themselves. And no study evaluated the uniqueness of laughter's physiological profile by contrasting it with other energetic but arbitrary vocalizations like shouting or cheering. It's time to congratulate the pioneers, learn from their efforts, and start over using better experimental designs and procedures. But something more needs to be said about this research because of its high profile in the popular media and because it bears on the effect of laughter on immune function, pain reduction, and health, topics to be considered later.

One of the best-known studies of laughter's biochemical correlates was conducted by Lee Berk and colleagues, who analyzed a series of blood samples collected from subjects watching a comedy video. This serial blood sampling technique can reveal rapid and fleeting changes in hormone levels that are beyond the resolution of the earlier urine work. (Corticotropin [ACTH], cortisol, beta-endorphin, dopac, epinephrine, norepinephrine, growth hormone, and prolactin were measured.) It was surprising that this sensitive analysis detected a laugh-related *reduction* in catecholamines and other correlates of sympathetic activation, a finding *opposite* that of previous research. Despite the impressive array of biochemical measures and sophisticated blood-sampling procedure, these data are difficult to interpret. Berk's widely publicized study of the biochemical effects of "laughter" is based on only five subjects (five additional subjects are in the control condition) and never reported whether any subjects actually laughed. The analysis is weakened further by having no true control condition—the control subjects didn't watch a neutral control video on knitting, or a drama, but spent "quiet time" in a room engaged in unspecified activities. The experimental effect of this laughterless study of "laughter" may only be about some unidentified effect of video watching.

After considering laughter's uncertain physiological correlates, we now pursue a tantalizing and even more challenging question. Does laughter, a sense of humor, or a lighthearted personality improve your health, add years to your life, prevent the common cold, or alleviate pain? We can roughly estimate the health benefits of these factors by correlating them with disease and longevity, but such approaches are difficult and fraught with possible confounds—healthy people are likely to be happy, and vice versa. And it's hard to link cause and effect over days, weeks, or years, as such studies require. In scrounging for the scarce and scattered data on health and humor, we broaden our search beyond laughter to include the more general variables of sense of humor, positive life events, cheerfulness, optimism, and other empirical odds and ends.

The immune system, one of our body's important defenses against disease, is the target of preliminary but promising research on humor's health effects. Because stress is associated with immunosuppression, it has been reasoned that humor may moderate such effects and increase immune-system function, and perhaps even health. However, this account is complicated because the presumed stress-reducing properties of humor are not well established and the vigorous *act* of laughter may itself produce transient stresslike physiological symptoms. If laughter reduces stress, it may be an aftereffect. Most immune system research examines salivary immunoglobulin A (S-IgA), an easy-to-study antibody that fights upper respiratory infections. The availability of kits to measure S-IgA levels prompted several psychologists to become instant immunologists, conducting a study or two before moving on to other problems. Most of these studies focused on the effects of *humor* and did not specifically examine laughter during their test conditions. Once again, the effect of humor is confounded with that of the powerful motor act of laughter.

Starting in 1985, a few small-scale studies begin to suggest that comedy and a person's ability to use humor to deal with daily events ("coping humor") boosted antibody (S-IgA) levels. It was already known that daily hassles (an index of stress) were associated with low S-IgA levels. Laughter and good humor promised a healthful antidote to life's miseries. In one of the most sophisticated studies relating life events to S-IgA immune function, Arthur Stone and his colleagues collected daily saliva samples and questionnaires from subjects over a 12-week period. On a day-to-day basis, high saliva antibody levels were associated with positive leisure and domestic events, and low levels were associated with negative work-related incidents. The implications? For the highest antibody levels and perhaps the best health, hang around the house, play with friends and family—and avoid fights with your boss.

Humor and lifestyle effects on immunity may not be restricted to S-IgA. In three short, one-paragraph abstracts, Lee Berk and colleagues reported laughter-related increases in immune system function, including lymphocyte blastogenesis and natural killer-cell activity. This increased activity was attributed to the "eustress" (desirable stress) of laughter that is presumed to lift the immunosuppressive effects of stress hormones. Unfortunately, all three studies apparently used flawed methods identical to the previously considered stress hormone study by this group and are subject to the same problems—nonconfirmed laughter, small sample size, and an inadequate control condition. As with the previous work, the reported experimental effects may be the result of video watching, not laughter or humor. It's sobering to realize that these three abstracts are the basis of much of the folklore about laughter's positive effect on immune function.

Several nonimmunological studies have found no relation between humor and health. In one such study, Albert Porterfield used two questionnaires, one to access sense of humor, and a second to measure physical symptoms, to examine if humor confers better health to Oberlin College students. Based on these pen-and-paper results, students with high humor scores were no more healthy than their less mirthful classmates.

Does childhood personality predict longevity? This provocative and important question about the long-term effects of personality on health was pursued in a rigorous and intriguing, large-scale study of 1,178 males and females. Howard Friedman and his colleagues collected data from the famous seven-decade longitudinal study initiated by Louis Terman in 1921. Contrary to their expectation, the authors found cheerfulness (optimism and sense of humor) in childhood to be *inversely* related to longevity. In contrast, conscientiousness was related to survival in middle to old age. They suggest that to the extent that optimism and humor are healthy, they may act as adaptive coping mechanisms to a transient crisis, rather than lifelong predispositions. Untempered optimism may even be maladaptive—the same optimism that may help someone pull through an emergency may facilitate risky behavior such as smoking or reckless driving ("I'll be OK"). Optimistic people may also be especially shocked and stressed by life events that turn out worse than expected. The Panglossian lifestyle may have hidden costs. These provocative data force a reconceptualization of issues relating sense of humor, cheerfulness, optimism, and related personality traits to health.

George Burns lived to the uncommonly old age of 100, although his wife and partner, Gracie Allen (1906–1964), survived to age 58, and controversial comic Lenny Bruce (1926–1966) managed only 40 troubled years. Which is the more typical longevity among professional funny men and women? (Gracie Allen actually considered herself an actor—straight man to George Burns, the comedian.) If sense of humor confers good health, comics who live by their wit should survive very long indeed—younger comics waiting in the wings will be wondering when the Jackie Masons, Rodney Dangerfields, Milton Berles, and Alan Kings will finally totter off the stage and make room for the next generation.

In a clever study, James Rotton tested the hypothesis about long-lived comics by comparing the longevity of comedians with that of other entertainers and nonentertainers. Alas, comedians don't have the last laugh—on the average, they and other entertainers died at an earlier age than nonentertainers, perhaps due to the stresses of being on the road, staying up late to do shows, or subjecting themselves and their art to the ruthless judgments of audiences and business managers. However, the cliché of the sad or bitter clown was not confirmed. Comics are not significantly prone to suicide, a symptom of depression, or to be victims of homicide, the ultimate bad audience reaction.

There is little scientific support for the popular idea that people with the personality traits of humor, cheerfulness, or optimism are particularly healthy or long-lived, but the possibility remains that situational laughter and humor are effective coping mechanisms for transient stress. The health-sustaining factor may not be laughter itself but how laughter and humor are used to confront life's challenges. Rod Martin and Herbert Lefcourt explored this idea by showing subjects a stressful film of a bloody aboriginal initiation rite and asked them to describe it in a humorous style. Subjects best able to do this reported the least stress in their lives, a finding that leaves me wondering what else is special about this select group.

Michelle Newman measured the stress reduction provided by a specific humor intervention in response to a specific stressor. Newman had subjects view a highly stressful film about three grisly industrial accidents, which were narrated by subjects using either a humorous or serious style, while she monitored three measures of physiological stress: heart rate, skin conductance, and skin temperature. The humor production group had the lowest negative affect, tension, and psychophysiological reactivity. To be effective, the humorous narrative need not be funny; it's sufficient for the subjects simply to give it a try. Related techniques of humor production can be taught and learned as part of a stress reduction program.

Pain is most exquisite when it, through nagging insistence, totally captures our being. But experience suggests that blessed distractions lighten pain's fierce embrace, if only momentarily. There is anecdotal evidence of pain's highly complex and psychological nature, and the potential for mirthful, analgesic intervention. Case studies of good outcomes reviewed by Norman Cousins and others are at last gaining empirical support as pain reduction is emerging as one of laughter's promising applications.

The first systematic analysis of laughter's pain-reducing (analgesic) properties was conducted by Rosemary Cogan and colleagues. She contrasted the effects of one of four different conditions on the subsequently tested discomfort level produced by an overinflated blood-pressure cuff. The cuff was inflated up to the level that subjects reported discomfort. The four 20-minute conditions were: a laughter-inducing audio tape; a relaxation-inducing tape; a neutral narrative to control for distraction; or no tape at all. Unlike most studies of humor effects, only subjects who actually laughed were included in her sample. Subjects in the laughter (a Lily Tomlin performance) and relaxation (progressive muscle relaxation) conditions tolerated the highest levels of cuff discomfort. The discovery of a natural analgesic technique that requires no patient effort or training is encouraging.

Other investigators extended Cogan's discovery although they did not specifically examine the effects of laughter. Using Cogan's pressure cuff technique, Dolf Zillmann and his colleagues found that comedy (*A Night at the Improv*, *Married with Children*) increased the tolerance of discomfort experienced after viewing the films. As might be expected, the magnitude of the analgesic effect was less than that of Cogan's subjects, who had to laugh out loud as a condition for inclusion in her study. Unexpectedly, Zillmann found that tragedy (*Terms of Endearment*) but not serious drama (*L.A. Law*) also increased cuff tolerance. The intensity of emotional involvement and captivating quality of the stimuli may be the common denominator of these entertainments. Endocrinological research points to common physiological features of humor, tragedy, and sadness.

Instead of presenting the humor intervention *before* pain testing, as Cogan and Zillmann did, two studies *simultaneously* presented the pain and humor stimuli. Deborah Hudak and colleagues report that the tolerance of painful electrical stimulation was greater while viewing a comedy than a nonhumorous documentary video. Ofra Nevo and colleagues also document the analgesic property of comedy, adding that the funnier the perception of a humorous film, the longer subjects could endure the pain from their hand immersed in ice water. But Nevo's "humor" condition provided no more pain relief than a documentary, suggesting that the pain relief was actually due to distraction or a related cognitive-behavioral process, not specifically comedy. The weakness of a humor or other intervention in these conditions is hardly surprising. With a painfully aching hand in ice water (Nevo), or while enduring electric shock (Hudak), how entertaining (or analgesic) can Bill Cosby or any other comedian be? Intense pain will overwhelm more subtle competing stimuli.

James Rotton and Mark Shats bypassed the benign ordeals contrived by laboratory scientists and observed the effects of humor on the alleviation of the serious pain produced by actual physical trauma. In a study of orthopedic surgery patients, they found that viewing movies of any sort, humorous or serious, reduced reports of distress and pain between the first and second day after surgery. Although there was no difference between patients viewing serious and comedic films, the comedy group requested fewer "minor analgesics" (aspirins, tranquilizers) than the serious group.

There is a significant eye-of-the-beholder effect in comedy. If patients can choose the films they watch, comedy viewers require lower doses of major analgesics (Demerol, Dilaudid, Percodan) than those viewing serious films. (Although not tested, pornography may have similar analgesic effects for those who like it.) However, the desired comedy effect is lost if there is no choice—annoying comedy can increase analgesic use. Bad drama is boring—bad comedy is obnoxious. Rotton and Shats's findings have broad implications for the often inconsistent results of comedy research. If subjects hate the comedy film selected by the experimenter, there may be no humor condition at all.

Before moving on, we should hear from society's most notorious purveyors of anxiety and pain, the dentists. In a brief research note, Ashton Trice and Judith Price-Greathouse found that subjects who joked, laughed, and used humor to cope with difficult circumstances experienced less stress during dental surgery than less humorous patients.

The scientific record offers modest but growing support for the analgesic properties of humor and laughter. But several critical questions are left unanswered. Does laughter or humor mediate the effect? Can analgesia be rendered by any compelling stimulus from *Hamlet* to a gentle caress? Essential details about the underlying biological mechanism remain obscure. One possible explanation of analgesia involves the release of endorphin, the body's own opiate-like substance. However, the only attempt to detect laughter-produced endorphins was unsuccessful. No one has tried the complementary strategy of examining whether the opiate- and endorphin-blocking drug Naloxone negates the analgesic effect. The failure to detect physiological correlates of psychological states should be neither surprising nor discouraging at this early stage of research. Until laboratory procedures improve, and we learn more about what, when, and how to measure, *behavioral evidence may be our best and most sensitive guide to physiology.* However, the acceptance of humor/laughter interventions by the medical profession will ultimately depend on the discovery of the mechanism through which humor manifests its analgesic effects. This is a problem well suited for an interested amateur armed with a comedy video, a pain-rating scale, and a pressure cuff or bucket of ice water.

The role of humor in psychotherapy has been the subject of debate and a lot of essay writing, but not much empirical research, a situation typical of the now fading philosophical/literary tradition in psychiatry and psychoanalysis. Some of the most relevant research on the therapeutic aspects of humor is considered elsewhere in this chapter in the broader context of health benefits, including potentially useful risk analyses, clinical interventions, and coping strategies. Although humor may seem rather benign, not everyone believes it to be so, including such philosophical masters as Plato and Aristotle who long ago warned of laughter's dark side (see Chapter 2). Among contemporary psychotherapists, the issue is still contested, with the pro-humor forces apparently gaining the upper hand in our "Don't worry, be happy," "Have a nice day" society. However, to some therapists, humor is "contraindicated," as physicians like to say when recommending against an agent or procedure.

Psychiatrist Lawrence Kubie warns of the "destructive potential of humor in psychotherapy" in an anecdote- and opinion-based essay that is typical of the genre. His concern is not simply with such blatant transgressions as ridiculing a depressed patient about his woes, beating him about the head with a pig bladder slapstick-style, or chanting "loser, loser." He cautions more broadly about the use of *any* humor in psychotherapy, noting that he "cannot point to a single patient in whose treatment humor proved to be a safe, valuable, and necessary aid." Clearly, some forms of humor can be deleterious to therapy, such as when the work of a session is deflected by joke telling, the use of humor as a defensive technique by the client to shield matters needing discussion, or the witty intellectual preening or aggressive humor of the therapist. Although Kubie's reservations are legitimate and generally shared by many other humor critics and proponents, his cure seems worse than the disease. Proponents of humor in psychotherapy emphasize that levity can be a means of broaching sensitive topics, providing insight, defusing anger, and relating to the client in a more intimate way. Humor, like violin playing, can be an instrument of art or aggression depending on the taste, sensitivity, and skill of the performer.

Although we know little about the effect of humor or laughter in psychotherapy, we do know that the more general property of voice quality can alter the mood of both speaker and listener in remarkable ways. In developing this intriguing topic, I ask the reader's indulgence. As is often the case, the critical research concerns not laughter but the more studied negative emotions of anger, fear, and anxiety.

The emotional impact of hearing "I love you" or "This is a stick-up" is undeniable, but recent research by Aron Siegman and his colleagues demonstrates powerful psychological and physiological effects of *how* we say something, an aspect of the prosodic dimension of speech. In research targeting cardiovascular risk factors, they found that talking in a loud, rapid voice like an angry person increases blood pressure, heart rate, and feelings of anger in the speaker, especially when matters of an emotional nature are being discussed. In other words, speak like a Marine drill instructor and you will start to feel and act like one! The inner experience of anger, by itself, did not drive cardiovascular reactivity.

Notice how your own anger builds when next you describe loudly an anger-producing event. Recounting anger-producing past events in a soft and slow (anger-inconsistent) voice produces lower speaker blood pressure, heart rate, and feelings of anger than when describing events in a loud and fast (anger-consistent) voice. Applications of this research to mood control are easily implemented and require little, if any, training. To moderate the escalation of blood pressure, heart rate, and anger in emotionally turbulent situations, you should speak softly and slowly, avoiding the bellicose vocal style that is by itself sufficient to drive blood pressure and aggression. Conversely, to become more assertive, use a loud, fast voice.

Although Siegman has not yet extended his research to include the speaker's audience, the results can be anticipated. The speaker and audience are engaged in an emotional pas de deux modulated by the tone of the speaker's voice. So far, only the matching (congruence) of speaker and audience speech rate and loudness but not physiology has been examined. However, daily experience suggests that we respond to a loud, aggressive voice with cardiovascular and emotional reactions of our own, perhaps even barking back an angry rejoinder that further increases the arousal of everyone. Speaker and audience can become locked in an explosive, mutually reinforced escalation of physiology and emotion having unpleasant and perhaps grave consequences, including cardiovascular incidents and violence. If we can control our voices at such critical times, our physiology and other behaviors may follow. If we lack vocal control, there is much to recommend keeping our mouths shut. These points find unexpected support in contemporary music.

Gangsta rap is a raw and emotionally potent form of anger-consistent vocal music that effectively drives the physiology of both performer and audience. The fast, rapid-fire cadence and loud, in-your-face aggressive character of rap music (especially gangsta rap) is just the kind of vocalization known to arouse the speaker—use your own experience to judge its effect on listeners. The listener effect even acts remotely, a fact not lost on those sharing their art via boom boxes in public places. Certainly, some aspect of rap or the gangsta rap lifestyle is provoking audiences.

The emotional power of rap to attract, repel, delight, and offend is due more to the style of its vocal delivery than the often controversial (and semi-intelligible) words so dear to the culture police. Loud, pulsative rock-and-roll lyrics have a similar but less potent effect. Like it or not, rap works, and is one of the significant developments in music and poetry in recent decades. Rap lyrics gain priority access to the emotional centers of the brain—they kick in the doors of our auditory attention centers and demand to be heard.

The voice of anger in rap or rant is a potent standard against which to compare laughter. Although laughter, the paradoxical voice of both joy and derision, seems as vocally vigorous and emotionally loaded as the voice of anger, its cardiovascular consequences are largely unknown. We don't know if recounting positive experiences in an upbeat manner accompanied by laughter will drive our physiology in ways opposite or different from anger.

The anger studies indicate the significance of vocalization in physiological reactivity, suggesting that the act of laughter, not the perception or production of humor, is the engine driving most physiological change. In ways presaged by psychologist William James, the voice may drive, or at least be an equal partner in, the production of the speaker's emotions—it is not a passive auditory signal of internal Sturm und Drang.

The evaluation of the health issues of laughter and humor now comes to a rather tentative close. As we review the varied leads, some false and some true, many complex and confusing, we witness the signs of a scientific and medical enterprise at its exciting early stages, when the most important work is yet to be done. We are describing traits of an unknown beast that is identified only through its shadow and footprints. We are one step removed from the beast itself, the underlying mechanism that links laughter and humor with such psychophysiological correlates as heart rate, blood pressure, immune function, analgesia, and ultimately mood and health. With the discovery of this mechanism will come a clarity and understanding not present in today's scattered correlational evidence—at last we will be able to emerge from the shadow world of psychophysiology and psychometrics and study real behavior and its neural basis.

Research on medicinal laughter, like many other promising enterprises (e.g., genetic engineering, artificial intelligence, the Internet), will pay a price for the burst of early exuberance with a backlash of undue pessimism before rebounding to a more realistic level. In the wake of overly optimistic predictions in the style of Norman Cousins and other proponents of holistic medicine, we are probably entering a downward phase, as disillusioned investigators realize that the necessary science is neither as easy nor as obvious as first anticipated, and negative results will start accumulating, often unpublished, in their filing cabinets. But the present mixed results hardly deserve the scathing editorial of Marcia Angell in the *New England Journal of Medicine* (1985) that notes that the current evidence for mental states' affecting the cause or cure of disease is largely folklore, little better than that of earlier centuries.

Given our present state of knowledge, laughter would deservedly be banned from the marketplace if it were a drug with unknown side effects. However, the advantageous cost/benefit ratio of laughter is such that there is no need to await FDA approval before giving laugh programs a try. (Suggestions about how to increase laughter in your life are provided in Chapter 3 and the Appendix.) The potential downside of laughter is small indeed, particularly if elected by patients. The promise of improved mood and quality of life without notable negative side effects is reason enough to implement experimental laughter or humor programs in health-care settings, even if welcome relief is provided only through placebo or distraction—there are worse outcomes than providing entertainment to patients desperately in need of it. Faster and better physical healing through laughter remains an unrealized, tantalizing, but still reasonable prospect.

APPENDIX

Ten Tips for Increasing Laughter

Perspectives from the Mall, Workplace, and Clinic

Ten years of prospecting taught me where to look for laughter. After the early days of trial and error, I detected underlying patterns and became more adept at predicting the social and physical settings most likely to yield laughter. Along the way, I acquired another useful insight—how to increase the laughter in our lives. Although I'm not a clinical psychologist and don't run Doctor Feelgood, Inc., it's clear that laugh enhancement is a useful enterprise. Whether the setting is commercial or therapeutic, laughter is generally associated with good times, good performance, improved attitude, and desirable outcomes. Based on my successes and failures to find laughter in the community, I offer the following 10 tips for increasing laughter. The tips focus on general principles, not specific techniques. These suggestions are informed predictions and have not been tested empirically. They are, however, a good starting point for people designing programs or environments for increasing laughter. No advice is offered about laugh suppression. It's easy to kill laughter, and many people and enterprises already have mastered this more modest art.

1. *Find a friend or personable stranger.* First and foremost, laughter is a social signal that almost disappears in solitary individuals. Happily, the critical social mass to produce laughter is only two—a pair of compatible people is enough to get laughter underway. If a person is unable or unwilling to seek social contact, visitors, especially past laugh-mates (friends), can provide the stimulation essential for laughter. For recluses, hospital patients, invalids, or nursing home residents, the television set is better than nothing, offering vicarious but noninteractive social contact—a window to a now distant social world. But another person is a much more effective laugh stimulus than a television set—and harder to ignore or turn off. Laughter is an arena in which companion animals such as dogs and cats fall short. However cuddly, comforting, affectionate, and otherwise wonderful our pets may be, they don't stimulate as much laughter (or as much anger) as other humans. Pets usually lack the critical social codes necessary to unlock our laugh response.

2. *The more the merrier.* Other things being equal, whether at a cocktail party or the cinema, a large crowd laughs more than a small one. Comedians know how difficult it is to milk laughs from a nearly empty house. The social multiplier effect brightens even the mood of customers at malls, the venue of much of my research. The mall is our modern-day, climate-controlled, main-street-in-a-box. Visitors to a bustling, successful mall come to shop, but once there, may stay to hang out with friends, see and be seen, and, to merchants' delight, make impulse purchases. Contrast this lively scenario with that of a desolate, laughless, dying mall where mission-oriented customers dart in to make a purchase and then escape to their waiting cars. The prescription for increasing laughter in a shopping center is similar to that for increasing sales and customer volume. Laughter provides a good, though indirect, measure of the success of a retail enterprise.

3. *Increase interpersonal contact.* Increasing face-to-face, eye-to-eye contact between group members maximizes laughter. At a party, a good host introduces guests to each other so that they aren't socially stranded (and laughless). Bigger groups do not necessarily increase laughter if group members are isolated from each other. For example, friends whisking down a mall concourse en route to a shop or the food court are largely laughless until they reach their destination and turn to face each other. Most of my observations of laughter were made at these sites of high interpersonal contact.

4. *Create a casual atmosphere.* Time-pressured, harried people don't laugh much—worry and anxiety kill laughter. A casual environment must also be safe—how relaxed will you feel at the mall if eyed by a group of surly adolescents who seem curious about the contents of your handbag or where you parked your car? Conventioneers provide a high laugh-yield because they have escaped temporarily from the stresses of home and office. (They also benefit from other factors noted above.) Malls and shops often create the leisurely atmosphere ideal for both selling and laughter by providing a comfortable environment and by removing clocks and other cues of urgency. (Try to find a clock in a shop.) New-style bookstores now offer espresso bars, live music, readings, and other innovations to make themselves into a site for mixing and meeting with other members of the bookish set. The mall-like settings of modern airports with their shops and fast-food restaurants offer an informative counter-example. Stress, urgency, fatigue, fear of flying, and sadness over impending separations drain laughter even from traveling friends and family. A reunion at the end of the trip provides a laugh-filled exception to the emotionally barren airport environment. Airports offer a case where large

crowds are not associated with laughter—weary, anonymous travelers avoid eye contact and interaction as they briskly stride toward their departure gates. Even busy airports have the look and social atmosphere of dying malls. The upside of airports? You can read or work undisturbed, as long as you don't miss your flight.

5. *Adopt a laugh-ready attitude.* Although you probably can't produce convincing voluntary laughter, you can voluntarily choose to laugh more by lowering your threshold for laughter. Just be willing and prepared to laugh. (Conversely, you can choose to laugh less by raising your threshold.) Of course, a playful, laugh-ready attitude can be fostered by an appropriate social and physical environment. You attend a friend's party or a comedy performance with the expectation that you will laugh, a self-fulfilling prophecy. A meeting with the IRS brings a different expectation.

6. *Exploit the contagious laugh effect.* If you are among a group of laughing people, you are likely to join in with your own yuks. Consider that laughter causes laughter (Chapter 7). If you want to laugh more, seek out cheerful, good-natured people. (Conversely, avoid those with foul dispositions.) Our lives as social beings are filled with an endless, mindless, and highly effective laugh track. The contagious-laugh effect is one of the reasons why an increase in group size facilitates laughter—large groups are more likely to have laughing people to prime and trigger the laugh response of others. It's bizarre but true that the disembodied laughter of a laugh box can emulate a laughing person. The underlying contagious laugh effect provides an informative glimpse behind the veil of rationality and fiction of reality that scripts our lives.

7. *Provide humorous materials.* A recurrent theme of this book is that laughter is more often a consequence of relationships than of jokes. Of course, humorous materials in the form of jokes, cartoons, books, videos, films, and audio recordings are potent and reliable stimuli for laughter, a point so obvious that it hardly earns

the term "tip." But several subtleties allow us to advance beyond posting Dilbert cartoons on the office bulletin board and telling jokes during coffee breaks. When selecting humor for others, it's important to appreciate the variety of tastes and avoid one-size-fits-all comedy. Bad drama is boring—bad comedy is obnoxious. Many hospitals provide "humor carts" full of funny magazines, videos, joke books, photographs, toys, and other entertaining material for use by patients. The carts are a great idea, cost little, and provide the needed variety of humorous materials. (The most important part of the humor cart, though, is the person pushing it.) As an antidote to your own dark nights of the soul, preemptively assemble a shelf of your favorite humorous cartoons, jokes, books, photographs, and videos—if you wait until you are depressed, you will probably not have the insight or initiative to take action.

8. *Remove social inhibitions.* The two approaches offered here are simple and not mutually exclusive. A person may increase laughter by either leaving an inhibiting setting or by personally becoming less inhibited. Members of fraternal organizations escape the critical eyes of the uninitiated, joining like-minded brethren in lodge houses where they perform strange rituals, wear silly hats, and do other things that may seem a bit odd to heathens in the outside world. Even the nonjoiners among us are less inhibited and laugh more when socially sheltered by accepting family and friends. Breezy, gossipy chats with colleagues are likely to be more laugh-filled if you can find a private space away from the prying eyes of co-workers. Although intimacy is hard to achieve while visiting a friend in a shared hospital room or ward, screens or curtains can be arranged to provide more privacy for conversation and laughter. (In some cases it may be appropriate to invite roommates to join in, reducing social inhibition by expanding the group size, and tapping social facilitation effects.) Playful group activities such as company sporting events (volleyball, etc.) may help break down barriers to communication and cure petty resentments acquired in other settings. (The more passive social encounters of

parties and picnics are less effective.) Alcohol and recreational drugs are deservedly celebrated and notorious for their ability to reduce social inhibitions and promote laughter. Happy hour is appropriately named, although alcohol and other pharmacological social lubricants bring the risk of physiological dependence and a loss of confidence in being able to function socially while sober. Some corporate humor workshops employ silliness exercises to reduce the social inhibitions of participants, teach humor skills, increase confidence and creativity, and perhaps cure the common cold. Certainly, there are merits to lightening the social tone of the workplace and getting colleagues to laugh with and at each other. But I can't escape the vision of brave corporate warriors shuffling out of a humor workshop and heading back to their cubicles to engage in capitalist warfare.

9. *Stage social events.* Social happenings bring people together, maximizing many of the already-mentioned factors driving laughter. Whether a gathering of friends, the company picnic, or the Annual Hog Festival, most folks enjoy an excuse to cheer, laugh, and revel. Even contrived social occasions are legitimized if the guests have a good time, a fact that should encourage potential organizers. The city of New Orleans successfully specializes in the human tendency to celebrate almost anything. Most efforts by retail establishments to generate social happenings are embarrassingly inept ("The Annual Toyota Sell-a-thon," "Joe's Mattress Extravaganza," etc.), but there are exceptions. "Retail theater" attempts to both sell and entertain. At Baltimore's Harbor Place mall, the Fudge Factory has employees who flamboyantly make fudge while dispensing comedy and song to potential customers who pause to watch the show. I don't know if they sell much fudge, but they do make a positive contribution to the mall experience and other stores will benefit. Other shops sell interesting stuff that visitors like to play with, such as the bizarre headwear at the Magic Hat, or offer special atmosphere like the Nature Store. "Mall theater" also includes the often limp offerings of musical groups, clowns, jugglers, showy sales demonstrations, Santa Claus, the Easter Bunny, merry-go-rounds, miniature trains, or the more sedate fare of visiting antique vendors and hobbyists, who set up in the mall concourses. Theme restaurants also provide their own brand of theater, with food often an afterthought. Consider Hard Rock Cafe, Rain Forest Cafe, Harley Davidson Cafe, and McDonald's, the granddaddy of all theme restaurants, with Ronald McDonald, Happy Meals, and play areas for kids.

10. *Tickle.* Last on the list is the most potent, ancient, and controversial laugh-stimulus. Like a powerful drug, tickle should come with a warning label and be used with care, because it can have harmful side effects. But when used as directed on the appropriate people of the right age, tickle is a part of physical play that brings satisfying consequences and laughter to both ticker and ticklee. First the warning. Tickle is the most intimate and difficult to ignore of all laugh-stimuli. In the language of touch, tickle is shouting. As experience has probably taught you, the pleasure of tickle is determined by who's doing it. Tickle is most appropriate for children, close family, friends, and lovers—never strangers. Indiscriminate tickling brings social and physical risk. Before commencing with this potent but risky laying on of hands, consult Chapter 6 for guidance.

NOTES

Chapter 1. Laughter: An Introduction

4 Motor development in birds: Provine, 1973; 1984.
5 Yawning: Provine, 1986; 1996a; 1997b.
8 Language-oriented books: Levelt, 1989; Handel, 1989.
9 Spemann, 1938, p. 367.

Chapter 2. The Road Not Taken

13 Negative consequences: Plato, *Republic,* pp. 84–86.
13 What makes a person laughable: Plato, *Philebus,* pp. 45–51.
13 Wit was a form of educated insolence: Aristotle, *Rhetoric,* p. 123.
13 Comic mask is distorted: Aristotle, *Poetics,* p. 229.
14 Those who go into excess: Aristotle, *Nicomachean Ethics,* p. 1000.
14 Gorgias: Aristotle, *Rhetoric,* p. 216.
14 Sudden glory: Hobbes, 1650/1999, pp. 54–55. Hobbes restates the sudden glory idea in a similar passage in *Leviathan,* 1651/1998, p. 38.
15 Laughter is an affection: Kant, 1790/1972, p. 177.
15 Laughter arises from the perceived mismatch: Schopenhauer, 1819/1907, p. 76.
15 Relief theory: Freud, 1905/1976.
16 Ragging: Bergson, 1911, pp. 135–136.
16 Absence of feeling: ibid, p. 4.
16 Must be human: ibid, p. 3.
17 Something mechanical encrusted upon the living: ibid, p. 37.
17 If we laugh each time: Koestler, 1964, p. 47.

17 If I find a bowling ball: Morreall, 1983, p. 64.
18 Tickle: Hall and Allin, 1897.
18 Introspective analyses: Martin, 1905.
18 Memory for funny material: Heim, 1936.
18 Laugh-provoking stimuli: Kamboropoulou, 1930.
18 Children's laughter: Kenderdine, 1931; Ding and Jersild, 1932; Justin, 1932.
18 Development: Darwin, 1872/1965; Washburn, 1929; Wilson, 1931.
19 Neurology: Wilson, 1924.
19 Psychiatry: Kraepelin in Defendorf, 1904; Bleuler, 1911/1950; 1916/1924.
19 Books that assembled: Chapman and Foot, 1976; McGhee and Goldstein, 1983a, b.
19 Why one person likes crude jokes: Eysenk, 1942.
20 Humor style and locus of control: Lefcourt, Sordoni, and Sordoni, 1974.
20 Incongruity: Rothbart, 1976.
20 Laugh more when bad things happen to obnoxious: Zillmann and Bryant, 1980.

Chapter 3. Natural History of Laughter

26 Who is actually laughing and why: Transcript analyses, most notably by Jefferson (1985) and Jefferson, Sacks, and Schegloff (1987), provide a microanalysis of laughter's placement in conversation. Although rigorous at the level of the transcript, these studies often make assumptions about the intentionality of laughter (e.g., laughter is "offered," "invited," or "accepted") that are unwarranted given laughter's minimal consciousness control.
26 Laugh episode: Provine, 1993.
26 Speaker and audience laughter: ibid. Glen (1989) examined the order of shared laughter in transcripts of conversations. The speaker laughed first in two-party interactions; In multi-party interactions, an audience member usually laughed first.
27 Gender differences in conversational styles: Tannen, 1990.
27 Gender differences in laughter: Provine, 1993.
27 Speakers laughed more: Provine, 1993.
27 More striking when gender was considered: The importance of considering the differential contribution and gender of speaker and audience to group laughter is emphasized by contrasting the present findings with those of Grammar and Eibl-Eibesfeldt (1990). Despite some methodological differences, both studies found more laughter from females than from males in mixed-sex groups, and neither study detected a difference between the *overall* frequency

of same-sex laughter at the level of the *group*. However, because Grammar and Eibl-Eibesfeldt did not distinguish between speaker and audience laughter, they did not detect the large gender difference in same-sex laughter (S_m=75.6 percent, A_m=60.0 percent; S_f=86.0 percent, A_f=49.8 percent). Averaging the laughter of female speakers and audiences cancels out the high level of speaker and low level of audience laughter.

28 Females are the leading laughers: Also see Adams and Kirkevold, 1978; Duncan and Fiske, 1977.

28 Male superiority in laugh getting: Foot and Chapman, 1976.

29 Cross-cultural study: Castell and Goldstein, 1977.

29 Target of humor and professional status: Coser, 1960.

30 Self-humbling: Brown and Levinson, 1978, p. 191. This chapter provides a rich variety of such behavior.

33 Men seek physical attractiveness and offer financial resources: Buss, 1989; Harrison and Saeed, 1977.

35 Young German adults: Grammar, 1990; Grammar and Eibl-Eibesfeldt, 1990.

36 Laughter punctuates speech: Provine, 1993.

37 Punctuation effect: ibid. In their study of the integration of laughter and speech in infant-directed speech of mothers, Nwokah, Hsu, Davies, and Fogel (1999) detected neither a punctuation effect nor speech dominance, a startling result given the strength of these effects. However, their apparent refutation included what I term *laughspeak*, a form of laughing speech excluded from my analysis of ha-ha-type laughs. They did not consider phrase interruptions by laughter, the basis of the present analysis.

39 The problem is that I can't follow the rhythm: Grandin, 1995, pp. 91–92.

40 What people say before they laugh: Provine, 1993.

42 Most laughter is not a response to jokes: ibid.

43 Do you really laugh when you are alone: Kierkegaard, 1843/1959, pp. 331–332.

44 Social setting essential for laughter: Provine and Fischer, 1989.

45 Awesome sociality: These data based on the *probability* (not frequency) of an act per hour are effective in revealing circadian and social trends, but are limited in their ability to provide direct comparisons between the relative frequencies of behaviors (i.e., 1 or 10 performances of an act during a given hour would yield identical probabilities of 1.0). Talking, for example, is much more than four times as frequent in social as in solitary situations.

45 Bowlers rarely smiled: Kraut and Johnston, 1979.

46 Olympic gold medal winners: Fernandez-Dols and Ruiz-Belda, 1995.

46 Babies at play tend not to smile: Jones and Raag, 1989.

46 Small talk may have evolved: Provine and Fischer, 1989, p. 304.
46 Grooming, gossip, and the evolution of language: Dunbar, 1996.
46 Small talk serves a big purpose: Tannen, 1990, p. 102.
47 Phatic speech: Malinowski in Farb, 1974, p. 23.
49 Tacit assumption of intentionality: Skinner (1986) gets it right, suggesting that laughter is an evolved, species-typical vocalization that can be simulated, as when laughing politely at an unfunny joke. Deacon (1997) provides a masterful analysis of those issues in the context of the co-evolution of language and the brain.
52 Voluntary (false) smiles are subtly different: Ekman and Friesen, 1982.
53 A study of split-brained patient P. S.: Gazzaniga and LeDoux, 1978, p. 142.

Chapter 4. Cracking the Laugh Code: From Sound Lab to Opera Studio

56 Sound description of the pretechnological age: In a reprinted classic, Mathews (1967) musically notates bird song.
57 The distinct acoustic signature: Provine and Yong, 1991.
59 Laughter as revealed in the sound spectra: Bachorowski, Smoski, and Owren (in preparation) review the confusing and often inconsistent earlier descriptions of laughter.
63 Although laughter is stereotyped: Bachorowski, Smoski, and Owren (in preparation) consider the variation of laughter with social context.
64 Laughter in congenitally deaf and blind children: Freeman, 1964; Eibl-Eibesfeldt, 1973.
64 Laughter may be emergenic: Lykken et al., 1992.
64 Giggle twins: ibid. Cherkas et al., 2000, in a study contrasting female monozygous and dizygous twins, report that "shared environment and random environmental factors, but not genetic effects," are responsible for humor appreciation, p. 17.
71 Problem of establishing tempo: Randel, 1986, p. 623.

Chapter 5. Chimpanzee Laughter, Speech Evolution, and Paleohumorology

75 Do we reign alone: Aristotle, *Parts of Animals*, p. 281.
75 Uniquely human behavioral trait: Recent statements of the homocentric hypothesis range from the distinguished scientific journal *Nature*, where Fried et al. (1998) begin their paper "Speech and laughter are uniquely human," to the Sunday newspaper magazine *Parade* (18 October 1998, "Ask Marilyn," *Washington Post*, p. 14),

where Marilyn Vos Savant (listed in the *Guinness Book of World Records* for the "highest IQ") reports that "Man is the only animal that laughs."

75 If a young chimpanzee be tickled: Darwin, 1872/1965, p. 131–132.

76 Young Orangs, when tickled: ibid., p. 132; Mackinnon, 1974.

76 Tickling between gorillas: Fossey, 1983, p. 120; Gorilla laughter, Fossey, 1972; Marler and Tenaza, 1977.

76 Primate vocalization experts: Marler and Tenaza, 1977; Goodall, 1986, p. 130; Van Lawick-Goodall, 1968; Yerkes, 1943; Berntson, et al., 1989.

77 The chimps emitted: Provine and Bard, 1994.

77 Ah grunting: van Hoof, 1972. Van Hoof presents a somewhat different perspective of laughter evolution than that offered here, although he agrees that human laughter evolved from a chimpanzeelike play vocalization and associated "play face" (also known as the "relaxed open-mouthed display"). Based on comparative analyses of a variety of primates, van Hoof proposed that human laughter is compounded from discrete ancestral displays. In humans, the play face and associated vocalization converged with the smile, an evolutionary derivative of the once independent "silent bared-teeth display," a submissive gesture in many primates. He suggests that human smiling is a display of friendliness that can be performed by itself or merged with laughter, a gesture blending in varying degrees the signals of friendliness (nonhostility) and playfulness. Although laughter and smiling are often evoked by similar stimuli, and smiling is more readily evoked than laughter, laughter and smiling have different evolutionary histories and a smile should not be interpreted as a low amplitude laugh. Lockhard et al. (1977), provide additional evidence of the discrete function and evolutionary heritage of laughs and smiles.

81 Why chimpanzees can't talk: Provine, 1996b; Provine and Bard, 1995.

82 Structure of the tongue, larynx, and vocal tract: Lieberman, 1984, provides an excellent review.

83 The woeful speech competence: Hayes and Hayes, 1951; Hayes, 1951.

83 Breakthroughs in human/primate communication: Gardner and Gardner, 1969; Patterson, 1978; Premack, 1972; Rumbaugh, 1977; Savage-Rumbaugh and Levin, 1994.

87 Speech and bipedal locomotion: Provine, 1997a, 1999.

87 Running and breathing: Bramble and Carrier, 1983.

88 Spinal cord structure: MacLarnon, 1995.

88 Fossil remains of Nariokotome *Homo erectus:* MacLarnon, 1993.

90 The African gray parrot: Pepperberg, 1987.

91 Hoover the seal: Deacon, 1997, reviews Hoover's vocal feats and their significance.

92 Tickling their own feet: Temerlin, 1975, also observed self-tickle in chimpanzee Lucy, p. 91.
93 Chimps sometimes laugh during solitary play: Roger Fouts, personal communication, 1998.
93 In a rare field study: Plooij, 1979.
94 Most candidates for simian humor: Roger Fouts, personal communication, 1998; Jensvold and Fouts, 1993; See McGhee, 1979, pp. 110–120, for additional anecdotes from Roger Fouts, Penny Patterson, and Duane Rumbaugh.
95 Dirty cat: Scatological misnaming by chimps Lucy and Washoe in Fouts, 1997, p. 157.
96 Bewildered by similar misnaming: McGhee, 1979, pp. 119–120.
96 Alcohol primes the laugh mechanism: Temerlin, 1975, p. 49.

Chapter 6. Tickle

100 Relegates tickle to chapter on pain: Sweet, 1959. Data relating pain, touch, and tickle are complex. Using cats, Zotterman (1939) found that nerves carrying pain information were also activated by light tactile (tickle?) stimulation. Human evidence comes from clinical cases in which the spinal pain pathways were transected to control intractable pain, a procedure that reduces (Laheurta, et al., 1990), but does not always eliminate (Nathan, 1990) the sensation of tickle.
100 Distinction between tickle and itch: Sweet, 1959, p. 499.
100 A young child, if tickled: Darwin, 1852/1965, p. 199.
101 Illegal in Norton, Virginia: Hyman, 1976, p. 102.
102 Being tickled was slightly less pleasant: in contrast to the gender difference in the hedonic quality of tickle, Harris and Christenfeld (1997, 1999) and Claxton (1975) detected no difference in the ticklishness of males and females, a result surprising to many self-proclaimed tickle-resistant males.
104 Tickle battles: Aldis, 1975, provides an insightful exploration of play fighting and many associated issues central to understanding laughter and play.
104 Populated with real tickle monsters: Arrowsmith, 1977; Ivantis, 1989; Warner, 1985.
104 It's more blessed to give than to receive: Chimpanzees also love to tickle as well as to be tickled, reports Fouts, p. 271.
107 A rare case of sadomasochism: Gutheil, 1947, p. 89.
111 Tickling them to death: Sade, 1797/1988, p. 797; Saint-Foix in Gould and Pyle, 1896/1956, p. 524.
111 Persecution of the Albigenses: Allin, 1903, p. 307.
111 Licked off by goats: Harris, 1999, reproduces a 1683 illustration, p. 345.

111 I always hated to be hugged: Grandin, 1995, p. 62.

112 Cerebellum to be abnormal: Bauman and Kemper, 1985.

113 Laughter develops: Sroufe and Waters, 1976; Sroufe and Wunsch, 1972.

114 Human primates at their most chimplike: Temerlin (1975) notes of chimpanzee Lucy, "Her favorite game is tickle-chase. Although she could easily outrun us, she almost never runs so fast that we cannot catch her to tickle her," p. 90. Researchers uniformly note chimpanzee's love of tickle and its power as a reinforcer.

116 Over 2,000 years ago: Aristotle, 1922.

116 Manually operated tickle machine: Weiskrantz, Elliot and Darlington, 1971.

116 A human's inability to tickle herself: Chimpanzees may have less cancellation of autostimulation. Temerlin (1975) reports of his home reared chimpanzee, "Lucy can send herself into peals of laughter by digging her fingers into her own muscles, usually those of the jaw, arms, or legs," p. 91.

117 Using a robotic tickler: Blakemore, Wolpert, and Frith, 1998.

119 Body-left and body-right function in relative independence: Provine and Westerman, 1979.

119 Can a machine tickle: Harris and Christenfeld, 1999. The papers of Harris (1999) and Harris and Christenfeld (1997, 1999), provide good, broad-based, reviews of the tickle literature.

119 Light tickle: Hall and Allin (1897) distinguish between light tickle (knismesis) and laughter-inducing heavy tickle (gargalesis).

120 Men even in a grieved state of mind: Bacon, 1677, p. 151.

120 Is tickle a reflex: The hybrid model offered in this chapter acknowledges both the strong neurologically based urge to respond to the stimulus (suggested by Harris [1999], Harris and Christenfeld [1999], and others as evidence of a "reflex") *and* the social context of tickle. By themselves, neither simple reflex- nor social-based explanations do justice to the complexity of the behavior, and such narrow interpretations are often based on anecdotes or archaic physiology. Typical *social* explanations argue that at least some aspect of tickle is grounded in intrapersonal experience (e.g., Darwin's already cited scenario). Foot and Chapman (1976) note that tickle's social context is essential to the production of laughter, p. 189. Other advocates of social determinants focus on the intent of the tickler: Koestler (1964) suggests that tickle evokes laughter only when it's interpreted as a harmless mock attack; Keith-Spiegel (1972) agrees that tickle must be administered by a "friendly" source," p. 18; and Shultz (1976) adds that "tickling must come from another person, otherwise it could not be interpreted as an attack," p. 32. The *reflex* position is supported by an equally diverse collection of evidence, the common denominator being the compelling

nature of the tickle stimulus and rejection of, or inattention to, social context. Although a variety of authors have described tickle as a reflex (Fridlund and Loftis [1990]; Hall and Allen [1897]; Sully [1902]; Stearns [1972]), their choice of mechanism may be due to convention or a lack of suitable alternative mechanisms.

122 A graph of body regions: Harris, 1999, p. 348.
124 The rather implausible idea: Darwin, 1872/1965; Heckler, 1873.
125 Indirect support for this proposition: Fridlund and Loftis, 1990.
125 Counter with an experimental study: Harris and Christenfeld, 1997.
125 Tickle is enhanced further by its conditionability: Newman, et al., 1993.
125 Can you tickle your dog or cat: Aldis, 1975, provides a review of animal play, laughter, and tickle, including the acts of play-biting, nibbling, and nuzzling.
126 Piles of writhing pachyderms: Joyce Pool, personal communication, 1998.
126 Rats responded: Panksepp and Burgdorf, 1999; Knutson, Burgdorf and Panksepp, 1998.
126 Physiology of play and joy: Panksepp, 1998.

Chapter 7. Contagious Laughter and the Brain

130 Contagious laughter in Tanganyika: Rankin and Philip, 1963. Also, Kagwa, 1964; Ebrahim, 1968; Muhangi, 1973.
132 Infectiousness of yawning: Provine, 1986, 1989a, 1989b.
132 Infectious crying: Simner, 1971.
132 Infectious coughing: Pennebaker, 1980.
132 Dance manias of the European Middle Ages: Rosen, 1968.
132 Mewing like a cat: Footnote by translator (Babington) in Hecker, 1846, p. 127.
132 Biting nuns: ibid.
132 Behavioral epidemics are still with us: Markush, 1973; Rosen, 1968.
133 Gallery of social exotica: Hatfield, Cacioppo, and Rapson, 1994, provide a wide-ranging review of emotional contagion.
134 Quakers actually quaked: Braithwaite, 1955, p. 57.
134 Going about naked: ibid., p. 148.
134 Shakers shook: Andrews, 1953.
134 Early Methodists: Rosen, 1968, p. 214.
134 America's unique religious heritage: Synan, 1971.
134 Cane Ridge: Weisberger, 1958, p. 35.
134 Azusa Street Mission: Synan, 1971.
135 Glossolalia (speaking in tongues): Malony and Lovekin, 1985.
135 Resurgence of holy laughter: Preston, 1994; Ostling, 1994. For a bliz-

zard of current information, both pro and con, about personalities, churches, and practices, do a Web search for "holy laughter." Followers of that old time religion have a burning passion for cyberspace.

138 Enthusiastically filled the void: For example: "He who laughs, lasts," *Saturday Review,* 25 April (1953), p. 30; "Can the laughs," *Saturday Review,* 6 March (1954), p. 28; "Strictly for laughs," *Newsweek,* 10 January (1955), p. 46; "Can the laughter," *Time,* 18 February (1957), pp. 38–40; "Laugh, soundman, laugh," *Saturday Review,* 18 April (1959), p. 44; "If its laughter you're after," *TV Guide,* 17 December (1960), pp. 6–7; "The Hollywood sphinx and his laff box," *TV Guide,* 2 July (1966), pp. 3–6; "Help! I'm a prisoner in a laff box," *TV Guide,* 9 July (1966), pp. 20–23; "Keyed for laughs," *Newsweek,* 16 December (1968), pp. 70–71; "The last laugh," *Newsweek,* 12 April (1971), p. 12; "Canned laughter—The absurd truth behind those snickers on TV," *Glamour,* October (1986), p. 110.

139 Roman emperor Nero: Griffin, 1985, p. 161.

139 The claque reached its fullest flower: *New Grove Dictionary of Opera,* v. 1, 1992, pp. 875–876; Crosten, 1948, pp. 41–48; Berlioz, 1963, especially pp. 93–110. Wechsberg, 1945, provides an insider's view of a more modern claque.

140 Rossini's *Il Barbiere di Siviglia* and Puccini's *Madame Butterfly:* Rosselli, 1984, p. 160.

140 According to Sir Rudolph Bing: Bing, 1972. Callas anecdotes, pp. 231–248.

140 Ed Wynn in a live radio broadcast: Bianculli, 1992, pp. 46–47.

147 As Kelly reported: Kelly, 1826, pp. 324–326.

147 Curmudgeon for all seasons: Chesterfield, 1774/1963 (letter of 19 October 1748), p. 84.

148 Laugh-box test: Provine, 1992.

149 Laugh-detector: ibid.

149 Sometimes baroque theorizing: ibid.

150 Neurological laugh-detector: ibid.; Provine, 1992, 1996a, b.

Chapter 8. Abnormal and Inappropriate Laughter

154 Pathological laughter: For reviews of the neurological literature, see Black, 1982; Brown, 1967; Davison and Kelman, 1939; Ironside, 1956; Martin, 1950; Poeck, 1969; Tilney and Morrison, 1912; Wilson, 1924.

155 Laughing death: Gajdusek and Zigas, 1959; Farquhar and Gajdusek, 1981; Zigas, 1990; Rhodes, 1987.

156 *Deadly Feasts:* Rhodes, 1987.

157 Masque manganique: Rodier, 1955; "This laughter spreads infectiously," ibid., p. 23.

157 *Odyssey:* Homer, 1967 (Book 20, lines 301–302), p. 306.

157 Except one weed: Pausanias, 1971, pp. 450–451. Translator Levi adds in note 115, "The plant was some sort of ranunculus; it tasted like a herb. There are 195 species of ranunculus in the Balkans, 65 of which are Mediterranean, so the lethal weed is hard to trace," p. 451.

158 Frozen, grinning mask: For a Gothic short story on this theme, see *Sardonicus* by Ray Russell, 1992, pp. 435–465.

158 Terrible and grotesque death: The sardonic smile provides a clue to Sherlock Holmes in *The Sign of Four,* Doyle, 1930.

158 Atmosphere of the highest possible heaven: Southey, in Erving, 1933, p. 5.

158 More unmingled pleasure: Coleridge, in Davy, 1800, p. 518.

158 Nitrous oxide was discovered: Davy, 1800; Wolfe and Menczer, 1994.

159 Young Samuel Colt: Rohan, 1935, pp. 28–37, 47–49.

159 P. T. Barnum: Raper, 1945, pp. 69–70.

159 Recreational effect of laughing gas that led to the discovery of anesthesia: Wolfe and Menczer, 1994; Keys, 1963.

160 Some dentists have become notorious: Willis, 1992.

160 Alcohol: Weaver, Masland, Kharazmi, and Zillmann, 1985. Alcohol can both facilitate laughter and change your sense of humor.

161 The effect of the gas: Colton, in Erving, 1933, p. 5.

161 *Child's Garden of Grass:* Margolis and Clorfene, 1969, pp. 33–34.

161 Intoxication progresses: From sense of subtle humor, Tart (1971), to smiling and giggling, Berke and Hernton (1974), to hilarity with minimal stimuli, Goode (1970).

161 Marijuana intoxication is not just a cultural quirk of the 1960s: Abel, 1980.

161 Marco Polo: Raulin, 1900. See Siegel and Hirschman, 1985, for entertaining historical notes and translations of early French investigators.

162 Egyptian cafes: Villard, 1872.

162 Club des Haschichins: Baudelaire, 1860/1971.

162 *Poem of Hashish:* ibid., p. 45.

162 Hashish and nitrous oxide laughter contrast: Raulin, 1900, p. 125.

162 He likes to laugh until tears come to his eyes and he is too weak to stand: Turnbull, 1961, p. 56.

162 When pygmies laugh: ibid., p. 44.

162 Cataplexy: Guilleminault, 1976.

163 H-reflex: Overeem, Lammers, and van Dijk, 1999.

163 Happy puppet/Angelman disorder: Elian, 1975; Williams and Frias, 1982; Hersh, et al., 1981.
163 A mother observed: Elian, 1975, pp. 903–904.
165 Chromosome 15: Nicholls, et al., 1992.
165 Seizure-produced laughter: Feeks, Murphy, and Porter, 1997.
166 Gelastic (laughing) epilepsy: Gascon and Lombroso, 1971; Chen and Forster, 1973; Arroyo, et al., 1993.
166 Involuntary crying/running: Chen and Forster, 1973; Luciano, Devinsky, and Perrine, 1993.
167 Seizure-prone 16-year-old girl: Fried, Wilson, MacDonald, and Behnke, 1998.
168 Children developing precociously: List, Dowman, Bagchi, and Bebin, 1958.
168 Reports from the land of ALS: Francis McGill, in Lieberman and Benson, 1977.
170 An ALS patient and her advocate: Personal communication, 1998
171 Great sage began acting queerly: de Santillana, 1961, p. 142.
172 Kraepelin's first descriptions of dementia praecox: Kraepelin, in Defendorf, 1904, pp. 167–168.
172 "Compulsive laughter" of schizophrenia: Bleuler, 1916/1924, p. 409.
172 A particularly frequent form of parathymia: Bleuler, 1911/1950, p. 52.
172 Usually the patients themselves: ibid., p. 452.
173 Laughing spells of four lobotomized schizophrenic patients: Kramer, 1954.
173 Auto-lobotomy: Colom and Levine, 1951.
176 Have you been playing with angels: The quotes in this paragraph were provided by parents of Rett daughters to the author, 1998.
176 Such laughter is symptomatic: Coleman, et al., 1988.
176 Williams disorder: Karmiloff-Smith, et al., 1995; Einfeld, Tonge, and Florio, 1997.
177 Wilson disease: Kolb and Whishaw, p. 299.
177 Alzheimer disease: Starkstein, et al., 1995; Kolb and Whishaw, 1990, pp. 824–833.
178 Pick disease: Cummins and Benson, 1983.
178 Tourette disorder: Kolb and Whishaw, 1990, pp. 300–302.
179 The President's Speech: Sacks, 1987, pp. 80–84.
180 Stroke damage on humor comprehension: Gardner, Ling, Flamm, and Silverman, 1975; Bihrle, Brownell, Powelson, and Gardner, 1986.
180 Joke-completion procedure: Brownell, Powelson, and Gardner, 1983.
182 Identifying the frontal lobe: Shammi and Stuss, 1999.
182 Galen considered joy to be more dangerous: Gould and Pyle, 1896/1956, p. 524.
182 A sample of such fatal moments: ibid.
183 Trollope had a stroke: Hall, 1991, p. 514.

183 American frontiersman: Hall and Allin, 1897, p. 7.
183 A woman of 58 was hospitalized: Andersen, 1936, in Martin, 1950, p. 457–458.
183 Uncontrollable laughter at graveside: Martin, 1950, p. 455–456.
184 Seinfeld Syncope: Cox, Eisenhauer, and Hreib, 1997.

Chapter 9. Laughing Your Way to Health

191 Anatomy of an illness: Cousins, 1976, 1979.
191 Moderated his laugh-your-way-to-health message: Cousins, 1989, pp. 126–127, 212, 214.
191 Earlier seminal work: Cannon, 1932; Selye, 1956.
192 Clown/physician Patch Adams: Adams, 1993.
192 Surgeon Bernie Siegel: Siegel, 1986.
192 Self-help guru Deepak Chopra: Chopra, 1989.
193 Work of laughter: Rowing anecdote from personal communication from Fry, 1991. Also see Fry and Rader, 1977.
193 Heart rate and blood pressure surge: Fry and Savin, 1988.
193 Depletes no more blood oxygen: Fry and Stoft, 1971.
193 Laughing Clubs International: Roach, 1996.
194 Biological movie review: Levi, 1965.
195 Investigators followed Levi's lead: Averill (1969) provided early, comprehensive contrast of the psychophysiological profiles of sadness and mirth. Sympathetic activation was common to both emotions, with cardiovascular changes being more prominent during sadness and respiratory changes more prominent during mirth. Further evidence of sympathetic activation during mirth comes from Schachter and Wheeler (1962), who found that preinjection of epinephrine increased the laughter of people viewing comedy. Although limited by idiosyncratic methods and presentation, Carruthers and Taggert (1973) reported a rise in epinephrine following a humorous section of a film, suggesting sympathetic activation, but also noted the apparently contradictory result of a slowing of the heart (bradycardia).
195 Investigators often disagree: Averill, 1969; Fry and Savin, 1988; Godkewitsch, 1976; Langevin and Day, 1972; and Jones and Harris, 1971, report that laughter and/or humor increase physiological arousal. Scheff and Bushnell, 1984; and Scheff and Scheele, 1980, report a decrease in arousal. White and Camarena, 1989, found no stress reduction effect of laughter, while White and Winzelberg, 1992, found laughter to be no more effective a stress reducer than relaxation. Analyses are complicated by the fact that laughter proba-

bly produces an initial arousal response, after which the measured variables may quickly drop to baseline or below. Future studies should track the dynamic changes during and after actual laughs and not focus on a probably unrepresentative psychophysiological snapshot taken some arbitrary time after the presumed humor/laughter intervention.

195 Laughter without humor: Fry and Savin, 1988, reported in passing that the blood pressure increase associated with faked laughs could not be differentiated from that accompanying mirthful laughter, suggesting that the motor act of laughter was sufficient to produce the effect.

195 Laughter's biochemical correlates: Berk, Tan, Fry, Napier, Lee, Hubbard, Lewis, and Eby, 1989.

197 A few small-scale studies: Dillon, Minchoff, and Baker, 1985; Martin and Dobbin, 1988; Dillon and Totten, 1989; Lefcourt, Davidson-Katz, and Kueneman, 1990. Labott, Ahleman, Wolever, and Martin, 1990, confirmed the immunoenhancing effect of humor, finding that the effect was not dependent upon overt laughter. In contrast, the immunosuppression of crying could be blocked if overt weeping was inhibited. Keeping a "stiff upper lip" may pay off with better immune system function.

197 Hassles (an index of stress) were associated with low S-IgA levels: McClelland, Alexander, and Marks, 1982; Kiecolt-Glaser, et al., 1987.

197 Relating life events to S-IgA immune function: Stone, et al., 1994.

197 Laughter-related increases in immune system function: Berk, et al., 1988; Berk, Tan, Napier, and Eby, 1989; Berk, Tan, Berk, and Eby, 1991.

198 No relation between humor and health: Porterfield, 1987.

198 Does childhood personality predict longevity: Friedman, et al., 1993.

199 Longevity of comedians: Rotton, 1992.

199 How laughter and humor are used to confront life's challenges: Martin and Lefcourt, 1983.

200 Specific humor intervention: Newman and Stone, 1996.

200 Laughter's pain-reducing (analgesic) properties: Cogan, Cogan, Waltz, and McCue, 1987.

201 Did not specifically examine the effects of laughter: Zillmann, Rockwell, Schweitzer, and Sundar, 1993.

201 Simultaneously presented the pain and humor stimuli: Hudak, Dale, Hudak, and DeGood, 1991.

201 Analgesic property of comedy: Nevo, Keinan, and Teshimovsky-Arditi, 1993.

202 Study of orthopedic surgery patients: Rotton and Shats, 1996.

202 Dentists: Trice and Price-Greathouse, 1986.

202 Only attempt to detect laughter-produced endorphins was unsuccessful: Berk, Tan, Fry, Napier, Lee, Hubbard, Lewis, and Eby, 1989.
203 Destructive potential of humor: Kubie, 1971.
203 Cannot point to a single patient: ibid., p. 865.
204 Proponents of humor in psychotherapy: Olsen, 1976; Ellis, 1977; Pierce, 1985. Salameh, 1983, provides a review.
204 Effects of how we say something: Siegman and Snow, 1997.
205 Matching of speaker and audience time-patterning: Feldstein, 1998.
207 Scathing editorial: Angell, 1985.

REFERENCES

Abel, E. L. (1980). *Marihuana, the First Twelve Thousand Years.* New York: Plenum Press.

Adams, P. (1993). *Gesundheit!* Rochester, VT: Healing Arts Press.

Adams, R.M., and B. Kirkenvold (1978). Looking, smiling, laughing, and moving in restaurants: sex and age differences. *Environmental Psychology and Nonverbal Behavior*, 3, 117–121.

Aldis, O. (1975). *Play Fighting.* New York: Academic Press.

Allin, A. (1903). On laughter (A review of Sully's "An essay on laughter"). *Psychological Review,* 10, 306–316.

Andrews, E. D. (1953). *The People Called Shakers: A Search for the Perfect Society.* New York: Oxford University Press.

Angell, M. (1985). Disease as a reflection of the psyche. *New England Journal of Medicine,* 312, 373–375.

Aristotle (1922). *An Aristotelian Theory of Comedy with an Adaptation of the* Poetics *and a Translation of the* Tractatus Coislinianus (L. Cooper, trans.). New York: Harcourt, Brace.

——— (1941). *Nicomachean Ethics* (W. D. Ross, trans.). In R. McKeon (ed.), *Basic Works of Aristotle.* New York: Random House.

——— (1961). *Parts of Animals* (A. L. Peck, trans.). Cambridge, MA: Harvard University Press.

——— (1984). *Poetics* (I. Bywater, trans.). New York: Modern Library.

——— (1984). *Rhetoric* (R. Roberts, trans.). New York: Modern Library.

Arrowsmith, N. (1977). *Field Guide to the Little People.* London: Macmillan.

Arroyo, S., R. P. Lesser, B. Gordon, S. Uematsu, J. Hart, P. Schwerdt, K. Andreasson, and R. S. Fisher (1993). Mirth, laughter and gelastic seizures. *Brain,* 116. 757–780.

Averill, J. R. (1969). Autonomic response patterns during sadness and mirth. *Psychophysiology,* 5, 399–414.

Backorowski, J.-A., M. Smoski, and M. J. Owren (in preparation). Acoustic variability in laughter is associated with social context.

Bacon, F. (1677). *Sylva Sylvarum: or A Natural History.* London: S. G. & B. Griffin.

Baudelaire, C. (1860/1971). *Artificial Paradise* (E. Fox, trans.). New York: Herder & Herder.

Bauman, M., and T. L. Kemper (1985). Histoanatomic observations of the brain in early infantile autism. *Neurology,* 85, 886–894.

Bergson, H. (1911). *Laughter: An Essay on the Meaning of the Comic* (C. Bereton and F. Rothwell, trans.). New York: Macmillan.

Berk, L. S., S. A. Tan, D. B. Berk, and W. C. Eby (1991). Immune system changes during humor associated laughter. *Clinical Research,* 39, 124A.

Berk, L. S., S. A. Tan, W. F. Fry, B. J. Napier, J. W. Lee, R. W. Hubbard, J. E. Lewis, and W. C. Eby (1989). Neuroendocrine and stress hormone changes during mirthful laughter. *The American Journal of the Medical Sciences,* 298, 390–396.

Berk, L. S., S. A. Tan, B. J. Napier, and W. C. Eby (1989). Eustress of mirthful laughter modifies natural killer cell activity. *Clinical Research,* 37, 115A.

Berk, L. S., S. A. Tan, S. Nehlsen-Cannarella, B. J. Napier, J. E. Lewis, J. W. Lee, and W. C. Eby (1988). Humor associated laughter decreases cortisol and increases spontaneous lymphocyte blastogenesis. *Clinical Research,* 36, 435A.

Berke, J., and C. C. Hernton (1974). *The Cannabis Experience.* London: Peter Owen.

Berlioz, H. (1963). *Evenings in the Orchestra* (C. R. Fortescue, trans.). Baltimore: Penguin.

Berntson, G. G., S. T. Boysen, H. R. Baker, and M. S. Torello (1989). Conspecific screams and laughter: Cardiac and behavioral reactions of infant chimpanzees. *Developmental Psychobiology,* 22, 771–787.

Bianculli, D. (1992). *Teleliteracy: Taking Television Seriously.* New York: Continuum.

Bihrle, A. M., H. H. Brownell, J. A. Powelson, and H. Gardner (1986). Comprehension of humorous and non-humorous material by left and right brain damaged patients. *Brain and Cognition,* 5, 399–411.

Bing, R. (1972). *5,000 Nights at the Opera.* New York: Popular Library.

Black D. W. (1984). Laughter. *Journal of Nervous and Mental Disease,* 170, 67–71.

Blakemore, S.-J., D. M. Wolpert, and C. D. Frith (1998). Central cancellation of self-produced tickle sensation. *Nature Neuroscience,* 1, 635–640.

Bleuler, E. P. (1911/1950). *Dementia Praecox.* New York: International Universities Press.

——— (1916/1924). *Textbook of Psychiatry* (A. A. Brill, trans.). New York: Macmillan.

Braithwaite, W. C. (1955). In H. C. Cadbury (ed.), *The Beginnings of Quakerism.* Cambridge: Cambridge University Press.

Bramble, D. M., and D. R. Currier (1983). Running and breathing in mammals. *Science,* 219, 251–256.

Brown, J. W. (1967). Physiology and phylogenesis of emotional expression. *Brain Research,* 5, 1–14.

Brown, P., and Levinson, St. (1978). Universals in language usage: Politeness phenomena. In E. N. Goody (ed.), *Questions and Politeness Strategies in Social Interaction* (pp. 56–290). Cambridge: Cambridge University Press.

Brownell, H. H., D. Michel, J. Powelson, and H. Gardner (1983). Surprise but not coherence: Sensitivity to verbal humour in right-hemisphere patients. *Brain and Language,* 18, 17–34.

Buss, D. M. (1989). Sex differences in human mate preferences: Evolutionary hypotheses tested in 37 cultures. *Behavioral and Brain Sciences,* 12, 1–49.

Cannon, W. (1932). *The Wisdom of the Body.* New York: Norton.

Carruthers, M., and P. Taggart (1973). Vagotonicity of violence: Biochemical and cardiac responses to violent films and television programmes. *British Medical Journal,* 3, 384–389.

Castell, P. J., and J. H. Goldstein (1977). Social occasions for joking: A cross-cultural study. In A. J. Chapman and H. C. Foot (eds.), *It's a Funny Thing, Humor* (pp. 193–197). Oxford: Pergamon Press.

Chapman, A. J., and H. C. Foot (eds.) (1976). *Humour and Laughter: Theory, Research, and Applications.* London: Wiley.

Chen, R.-C., and F. M. Forster (1973). Cursive epilepsy and gelastic epilepsy. *Neurology,* 23, 1019–1029.

Cherkas, L., F. Hochberg, A. J. MacGregor, H. Sneider, and T. D. Spector (2000). Happy families: A twin study of humour. *Twin Research,* 3, 17–22.

Chesterfield, Lord (1774/1963) *Lord Chesterfield's Letters to His Son.* New York: Dutton.

Chopra, D. (1989). *Quantum Healing.* New York: Bantam Books.

Claxton, G. (1975). Why can't we tickle ourselves? *Perceptual and Motor Skills,* 41, 335–338.

Cogan, R., D. Cogan, W. Waltz, and M. McCue (1987). Effects of laughter and relaxation on discomfort thresholds. *Journal of Behavioral Medicine,* 10, 139–144.

Coleman, M., J. Brubaker, K. Hunter, and G. Smith (1988). Rett syndrome: a survey of North American patients. *Journal of Mental Deficiency Research,* 32, 117–124.

Colom, G. A., and M. Levine, (1951). A case of self-inflicted lobotomy. *Journal of Nervous and Mental Disease,* 113, 430–436.

Coser, R. L. (1960). Laughter among colleagues: A study of the social functions of humor among the staff of a mental hospital. *Psychiatry*, 23, 81–95.

Cousins, N. (1976). Anatomy of an illness. *New England Journal of Medicine*, 295, 1458–1463.

——— (1979). *The Anatomy of an Illness as Perceived by the Patient.* New York: Norton.

——— (1989). *Head First: The Biology of Hope.* New York: Dutton.

Cox, S. V., A. C. Eisenhauer, and K. Hreib (1997). "Seinfield syncope." *Catheterization and Cardiovascular Diagnosis*, 42, 242.

Crosten, W. L. (1948). *French Grand Opera: An Art and a Business.* New York: King's Crown Press.

Cummins, J. L., and D. F. Benson (1983). *Demensia: A Clinical Approach.* Boston: Butterworth.

Darwin, C. (1872/1965). *The Expression of Emotions in Man and Animals.* Chicago: University of Chicago Press.

———. (1877). A biographical sketch of an infant. *Mind*, 2, 285–294.

Davison, C., and H. Kelman (1939). Pathologic laughing and crying. *Archives of Neurology and Psychiatry*, 42, 595–643.

Davy, H. (1800). *Researches, Chemical and Philosophical; Chiefly Concerning Nitrous Oxide.* London: Butterworths.

Deacon, T. W. (1997). *The Symbolic Species.* New York: Norton.

Defendorf, A. R. (1904). *Clinical Psychiatry: A Textbook Abstracted and Adapted from the 6th German Edition of Kraepelin's "Lehrbuch der Psychiatrie."* New York: Macmillan.

Dillon, K. M., B. Minchoff, and K. H. Baker (1985). Positive emotional states and enhancement of the immune system. *International Journal of Psychiatry in Medicine*, 15, 13–17.

Dillon, K. M., and M. C. Totten (1989). Psychological factors, immunocompetence, and health of breast-feeding mothers and their infants. *Journal of Genetic Psychology*, 150, 155–162.

Ding, G. F., and A. T. Jersild (1932). The study of the laughing and smiling of preschool children. *Journal of Genetic Psychology*, 40, 452–472.

Doyle, A. C. (1930). *The Sign of Four.* In *The Complete Sherlock Holmes, Vol. 1* (pp. 91–173). New York: Doubleday.

Dunbar, R. (1996). *Grooming, Gossip, and the Evolution of Language.* Cambridge, MA: Harvard University Press.

Duncan, S. D., and D. W. Fiske (1977). Face-to-face interactions. In *Research, Methods and Theory*. Hillsdale, N. J.: Erlbaum.

Ebrahim, G. J. (1968). Mass hysteria in school children. Notes on three outbreaks in East Africa. *Clinical Pediatrics*, 7, 437–438.

Eibl-Eisenfeldt, I. (1973). The expressive behavior of the deaf-and-blind born. In M. von Cranach and I. Vine (eds.), *Social Communication and Movement* (pp. 163–194). London: Academic Press.

Einfeld, S. L., B. J. Tonge, and T. Florio (1997). Behavioral and emotional disturbance in individuals with Williams syndrome. *American Journal of Mental Retardation*, 1, 45–53.

Ekman, P., and W. V. Friesen (1982). Felt, false, and miserable smiles. *Journal of Nonverbal Behavior*, 6, 238–252.

Elian, M. (1975). Fourteen happy puppets. *Clinical Pediatrics*, 14, 902–908.

Ellis, A. (1977). Fun as psychotherapy. *Rational Living*, 12, 2–6.

Erving, H. W. (1933). The discoverer of anaesthesia: Dr. Horace Wells of Hartford. *Yale Journal of Biology and Medicine*, 5, May 1933. Reprinted by the Tercentennial Commission of the State of Connecticut.

Eysenk, H. (1942). The appreciation of humor: An experimental and theoretical study. *British Journal of Psychology*, 32, 295–309.

Farb, P. (1974). *Word Play*. New York: Knopf.

Farquhar, J., and D. C. Gajdusek (1981). *Kuru: Early Letters and Fieldnotes from the Collection of D. Carlton Gajdusek*. New York: Raven Press.

Feeks, E. F., G. L. Murphy, and H. O. Porter (1997). Laughter in the cockpit: Gelastic seizures—A case report. *Aviation, Space, and Environmental Medicine*, 68, 66–68.

Feldstein, S. (1998). Some nonobvious consequences of monitoring time in conversations. In G. A. Barnet and M. T. Palmer (eds.), *Progress in Communication Sciences*, vol. 14 (pp. 163–190). Norwood, NJ: Ablex.

Fernandez-Dols, J. M., and M.-A. Ruiz-Belda (1995). Are smiles a sign of happiness? Gold medal winners at the Olympic Games. *Journal of Personality and Social Psychology*, 69, 1113–1119.

Foot, H. C., and A. J. Chapman (1976). The social responsiveness of young children in humorous situations. In A. J. Chapman and H. C. Foot (eds.), *Humor and Laughter: Theory, Research and Application* (pp. 141–175, 187–214). London: Wiley.

Fossey, D. (1972). Vocalizations of the mountain gorilla *(Gorilla beringei)*. *Animal Behaviour*, 20, 36–53.

——— (1983). *Gorillas in the Mist*. Boston: Houghton Mifflin.

Fouts, R. (1997). *Next of Kin*. New York: William Morrow.

Freedman, D. G. (1964). Smiling in blind infants and the issue of innate vs. acquired. *Journal of Child Psychology and Psychiatry*, 5, 171–184.

Freud, S. (1905/1976). *Jokes and Their Relation to the Unconscious* (J. Strachey, trans.). Harmondsworth, UK: Penguin.

Fridlund, A. J., and J. M. Loftis (1990). Relations between tickling and humorous laughter: Preliminary support for the Darwin-Hecker hypothesis. *Biological Psychology*, 30, 141–150.

Fried, I., C. L. Wilson, K. A. MacDonald, and E. J. Behnke (1998). Electric current stimulates laughter. *Nature*, 391, 650.

Friedman, H. S., J. S. Tucker, C. Tomlinson-Keasey, J. E. Schwartz, D. L. Wingard, and M. H. Criqui (1993). Does childhood personality predict longevity? *Journal of Personality and Social Psychology*, 65, 176–185.

Fry, W. F., and C. Rader (1977). The respiratory components of mirthful laughter. *The Journal of Biological Psychology*, 19, 39–50.

Fry, W. F., and W. M. Savin (1988). Mirthful laughter and blood pressure. *Humor*, 1, 49–62.

Fry, W. F., and P. E. Stroft (1971). Mirth and oxygen saturation levels of peripheral blood. *Psychotherapeutics and Psychosomatics*, 19, 76–84.

Gajdusek, D. C., and V. Zigas (1959). Kuru. *American Journal of Medicine*, 26, 442–469.

Gardner, H., P. K. Ling, L. Flamm, and J. Silverman (1975). Comprehension and appreciation of humorous material following brain damage. *Brain*, 98, 399–412.

Gardner, R. A., and B. T. Gardner (1969). Teaching sign language to a chimpanzee. *Science*, 165, 664–672.

Gascon, G. G., and C. T. Lombroso (1971). Epileptic (gelastic) laughter. *Epilepsia*, 12, 63–76.

Gazzaniga, M. S., and J. E. LeDoux (1978). *The Integrated Mind.* New York: Plenum Press.

Glenn, P. J. (1989). Initiating shared laughter in multi-party conversations. *Western Journal of Speech Communication*, 53, 127–149.

Godkewitsch, M. (1976). Physiological and verbal indices of arousal in rated humor. In A. Chapman and H. C. Foot (eds.), *Humor and Laughter: Theory, Research and Application.* London: Wiley.

Goodall, J. (1986). *The Chimpanzees of Gombe: Patterns of Behavior.* Cambridge, MA: Harvard University Press.

Goode, E. (1970). *The Marijuana Smokers.* New York: Basic Books.

Gould, G. M., and W. L. Pyle (1896/1956). *Anomalies and Curiosities of Medicine.* New York: Julian Press.

Grammar, K. (1990). Strangers meet: Laughter and non-verbal signs of interest in opposite-sex encounters. *Journal of Nonverbal Behavior*, 14, 209–236.

Grammar, K., and I. Eibl-Eibesfeldt (1990). The ritualization of laughter. In W. A. Koch (ed.), *Naturlichtkeit der Sprache und der Kulture. Bochumer Beitrage zur Semiotik* (pp. 192–214). Bochum, Germany: Brockmeyer.

Grandin, T. (1995). *Thinking in Pictures.* New York: Doubleday.

Griffin, M. T. (1985). *Nero: the End of a Dynasty.* New Haven: Yale University Press.

Guilleminault, C. (1976). Cataplexy. In C. Guilleminault, W. C. Dement, and P. Passouant (eds.), *Narcolepsy* (pp. 125–143). New York: S P Books.

Gutheil, E. A. (1947). A rare case of sadomasochism (Torture by tickling). *American Journal of Psychotherapy*, 1, 87–92.

Hall, J. N. (1991). *Trollope: A Biography.* Oxford: Clarendon Press.

Hall, G. S., and A. Allin (1897). The psychology of tickling, laughing and the comic. *American Journal of Psychology*, 9, 1–42.

Handel, S. (1989). *Listening: An Introduction to the Perception of Auditory Events.* Cambridge, MA: MIT Press.

Harris, C. R. (1999). The mystery of ticklish laughter. *American Scientist,* 87, 344–351.

Harris, C. R., and N. Christenfeld (1997). Humour, tickle, and the Darwin-Hecker hypothesis. *Cognition and Emotion,* 11, 103–110.

——— (1999). Can a machine tickle? *Psychonomic Bulletin & Review,* 6, 504–510.

Harrison, A. A., and L. Saeed (1977). Let's make a deal: An analysis of revelations and stipulations in lonely hearts advertisements. *Journal of Personality and Social Psychology,* 35, 257–264.

Hatfield, E., J. T. Cacioppo, and R. L. Rapson (1994). *Emotional Contagion.* Cambridge: Cambridge University Press.

Hayes, C. (1951). *The Ape in Our House.* New York: Harper & Row.

Hayes, K. J., and C. Hayes (1951). The intellectual development of a home-raised chimpanzee. *Proceedings of the American Philosophical Society,* 95, 105–109.

Hecker, E. (1873). *Die Physiologie und Psychologie des Lachen und des Komischen.* Berlin: F. Dummler.

Hecker, J. F. C. (1846). *The Epidemics of the Middle Ages* (B. G. Babington, trans.). London: George Woodfall and Sons.

Heim, A. (1936). An experiment on humor. *British Journal of Psychology,* 27, 148–161.

Hersh, J. H., A. S. Bloom, A. W. Zimmerman, N. D. Dinno, R. M. Greenstein, B. Weisskopf, and A. H. Reese (1981). Behavioral correlates in the happy puppet syndrome: A characteristic profile? *Developmental Medicine and Child Neurology,* 23, 792–800.

Hobbes, T. (1650/1999) Human nature. In *Human Nature and Decopore Politico.* Oxford: Oxford University Press.

——— (1651/1998). *Leviathan.* Oxford: Oxford University Press.

Homer (1967). *The Odyssey of Homer* (R. Lattimore, trans.) New York: Harper & Row.

Hoof, J.A.R.A.M. van (1972). A comparative approach to the phylogeny of laughter and smiling. In R. A. Hinde (ed.), *Non-verbal Communication* (pp. 209–241). Cambridge: Cambridge University Press.

Hudak, D. A., A. Dale, M. A. Hudak, and D. E. DeGood (1991). Effects of humorous stimuli and sense of humor on discomfort. *Psychological Reports,* 69, 779–786.

Hyman, D. (1976). *The Trinton Pickle Ordinance (and Other Bonehead Legislation).* Brattleboro, VT: The Stephen Greene Press.

Ironside, R. (1956). Disorders of laughter due to brain lesions. *Brain,* 79, 589–609.

Ivantis, L. J. (1989). *Russian Folk Belief.* Armonk, NY: M. E. Sharpe.

Jefferson, G. (1985). An exercise in the transcription and analysis of laughter. In T. A. van Dijk (ed.), *Handbook of Discourse Analysis, Vol. 3: Discourse and Dialogue* (pp. 25–34). London: Academic Press.

Jefferson, G., H. Sacks, and E. Schegloff (1987). Notes on laughter in the pursuit of intimacy. In G. Button and J. R. Lee (eds.), *Talk and Social Organization* (pp. 152–205). Clevedon, UK: Multilingual Matters Ltd.

Jensvold, M. L. A., and R. S. Fouts (1993). Imaginary play in chimpanzees *(Pan troglodytes)*. *Human Evolution,* 8, 217–227.

Jones, J. M., and P. E. Harris (1971). Psychophysiological correlates of cartoon appreciation. *Proceedings of the Annual Convention of the American Psychological Association,* 6, 381–382.

Jones, S. S., and T. Raag (1989). Smile production in older infants: The importance of a social recipient for the facial signal. *Child Development,* 60, 811–818.

Justin, F. (1932). A genetic study of laughter-provoking stimuli. *Child Development,* 3, 114–136.

Kagwa, B. H. (1964). The problem of mass hysteria in East Africa. *East African Journal of Medicine,* 41, 560–566.

Kambouropoulou, P. (1930). Individual differences in the sense of humor and their relation to temperamental differences. *Archives of Psychology,* 19, 5–77.

Kant, I. (1790/1972). *Critique of Judgement* (J. H. Bernhard, trans.). New York: Hafner.

Karmiloff-Smith, A., E. Kilma, U. Bellugi, J. Grant, and S. Baron-Cohen (1995). Is there a social module? Language, face processing, and theory of mind in individuals with Williams syndrome. *Journal of Cognitive Neuroscience,* 7, 196–208.

Keith-Spiegel, P. (1972). Early conceptions of humor: Varieties and issues. In J. H. Goldstein and P. E. McGhee (eds.), *The Psychology of Humor* (pp. 4–39). New York: Academic Press.

Kelly, M. (1826). *Reminiscences of Michael Kelly of the King's Theatre.* London: Henry Colburn.

Kenderdine, M. (1931). Laughter in the pre-school child. *Child Development,* 2, 228–230.

Keys, T. E. (1963). *The History of Surgical Anesthesia.* New York: Dover.

Kiecolt-Glaser, J. K., L. Fisher, P. Ogrocki, J. C. Stout, C. E. Speicher, and R. Glaser (1987). Marital quality, marital disruption, and immune function. *Psychosomatic Medicine,* 49, 13–34.

Kierkegaard, S. (1843/1959). *Either/Or. Vol. 2* (W. Lowrie, trans.). Garden City, NY: Anchor Books.

Knutson, B., J. Burgdorf, and J. Panksepp (1998). Anticipation of play elicits high-frequency ultrasonic vocalizations in young rats. *Journal of Comparative Psychology,* 112, 65–73.

Koestler, A. (1964). *The Act of Creation.* London: Hutchinson.

Kolb, B., and I. Q. Whishaw (1990). *Fundamentals of Human Neuropsychology.* New York: W. H. Freeman.

Kramer, H. C. (1954). Laughing spells in patients, after lobotomy. *Journal of Nervous and Mental Disease,* 119, 517–522.

Kraut, R. E., and R. E. Johnston (1979). Social and emotional messages of smiling: An ethological approach. *Journal of Personality and Social Psychology,* 37, 1539–1553.

Kubie, L. S. (1971). The destructive potential of humor in psychotherapy. *American Journal of Psychiatry,* 127, 861–866.

Labott, S. M., S. Ahleman, M. E. Wolever, and R. B. Martin (1990). The physiological and psychological effects of the expression and inhibition of emotion. *Behavioral Medicine,* 16, 182–188.

Lahuerta, J., D. Bowsher, J. Campbell, and S. Lipton (1990). Clinical and instrumental evaluation of sensory function before and after percutaneous anterolateral cordotomy at cervical level in man. *Pain,* 42, 23–30.

Langevin, R., and H. I. Day (1972). Physiological correlates of humor. In J. H. Goldstein and P. E. McGhee (eds.), *The Psychology of Humor* (129–142). New York: Academic Press.

Lefcourt, H. M., K. Davidson-Katz, and K. Kueneman (1990). Humor and immune system functioning. *Humor—International Journal of Humor Research,* 3, 305–321.

Lefcourt, H., C. Sordoni, and C. Sordoni (1974). Locus of control and the expression of humor. *Journal of Personality,* 42, 130–143.

Levelt, W. J. M. (1989). *Speaking: From Intention to Articulation.* Cambridge, MA: MIT Press.

Levi, L. (1965). The urinary output of adrenalin and noradrenalin during pleasant and unpleasant emotional states. *Psychosomatic Medicine,* 27, 80–85.

Levine, J. (1979). Humor and psychopathology. In C. Izard (ed.), *Emotions in Personality and Psychopathology* (pp. 37–69). New York: Plenum Press.

Lieberman, A., and F. Benson (1977). Control of emotional expression in pseudobulbar palsy. A personal experience. *Archives of Neurology,* 34, 717–719.

Lieberman, P. (1984). *The Biology and Evolution of Language.* Cambridge, MA: Harvard University Press.

List, C. F., C. E. Dowman, B. K. Bagchi, and J. Bebin (1958). Posterior hypothalamic hamartomas and gangliogliomas causing precocious puberty. *Neurology,* 8, 164–174.

Lloyd, E. L. (1938). The respiratory mechanism in laughter. *Journal of General Psychology,* 19, 179–189.

Lockard, J. S., C. E. Fahrenbruch, J. L. Smith, and C. J. Morgan (1977). Smiling and laughter: Different phyletic origins? *Bulletin of the Psychonomic Society,* 10, 183–186.

Luciano, D., O. Devinsky, and K. Perrine (1993). Crying seizures. *Neurology,* 43, 2113–2117.

Lykken, D. T., M. McGue, A. Tellegen, and T. J. Bouchard (1992). Emergenesis: Genetic traits that may not run in families. *American Psychologist,* 47, 1565–1577.

Mackinnon, J. (1974). The behavior and ecology of wild orang-utans *(Pongo pygmaeus). Animal Behaviour,* 22, 3–74.

MacLarnon, A. (1993). The vertebrate canal. In A. Walker and R. Leakey (eds.), *The Nariokotome Homo Erectus Skeleton* (pp. 359–390). Cambridge, MA: Harvard University Press.

——— (1995). The distribution of spinal cord tissues and motor adaptations in primates. *Journal of Human Evolution,* 29, 463–482.

Malony, H. N., and A. A. Lovekin (1985). *Glossolalia.* New York: Oxford University Press.

Margolis, J. S., and R. Clorfene (1969). *A Child's Garden of Grass. (The Official Handbook for Marijuana Users).* North Hollywood, CA: Contact Books.

Markush, R. E. (1973). Mental epidemics: A review of the old to prepare for the new. *Public Health Reviews,* vol. 2, no. 4, 353–442.

Marler, P., and R. Tenaza (1977). Signaling behavior of apes with special reference to vocalization. In T. A. Sebeok (ed.), *How Animals Communicate* (pp. 965–1033). Bloomington: Indiana University Press.

Martin, J. P. (1950). Fits of laughter (sham mirth) in organic cerebral disease. *Brain,* 73, 453–464.

Martin, L. J. (1905). Psychology of aesthetics: I. Experimental prospecting in the field of the comic. *American Journal of Psychology,* 16, 35–118.

Martin, R. A., and J. P. Dobbin (1988). Sense of humor, hassles and Immunoglobulin A-Evidence for a stress moderating effect of humor. *International Journal of Psychiatry in Medicine,* 18, 93–105.

Martin, R. A., and H. M. Lefcourt (1983). Sense of humor as a moderator of the relationship between stressors and moods. *Journal of Personality and Social Psychology,* 45, 1313–1324.

Mathews, F. S. (1967). *Field Book of Wild Bird Songs and Their Music.* New York: Dover.

McClelland, D. C., C. Alexander, and E. Marks (1982). The need for power, stress, immune function, and illness among male prisoners. *Journal of Abnormal Psychology,* 91, 61–70.

McGhee, P. E. (1979). *Humor: Its Origin and Development.* San Francisco: W. H. Freeman.

McGhee, P. E., and J. H. Goldstein (eds.) (1983a). *Handbook of Humor Research: Vol. 1. Basic Issues.* New York: Springer-Verlag.

——— (1983b). *Handbook of Humor Research: Vol. 2. Applied Studies.* New York: Springer-Verlag.

Morreall, J. (1983). *Taking Laughter Seriously.* Albany, NY: SUNY Press.

Muhangi, J. R. (1973). A preliminary report on mass hysteria in an Ankole school in Uganda. *East African Journal of Medicine,* 50, 304–309.

Muybridge, E. (1887/1957) *Animals in Motion*. New York: Dover.

——— (1901/1955) *Human Figure in Motion*. New York: Dover.

Nathan, P. W. (1990). Touch and surgical division of the anterior quadrant of the spinal cord. *Journal of Neurology, Neurosurgery and Psychiatry,* 53, 935–939.

Nevo, O., G. Keinan, and M. Teshimovsky-Arditi (1993). Humor and pain tolerance. *Humor,* 6, 71–88.

New Grove Dictionary of Opera (1992). Vol. 1. London: Macmillan Press Limited.

Newman, B., M. A. O'Grady, C. S. Ryan, and N. S. Hemmes (1993). Pavlovian conditioning of the tickle response of human subjects: Temporal and delay conditioning. *Perceptual and Motor Skills,* 77, 779–785.

Newman, M. G., and A. A. Stone (1996). Does humor moderate the effects of experimentally induced stress? *Annals of Behavioral Medicine,* 18, 101–109.

Nicholls, R. D., G. S. Pai, W. Gottlieb, and E. Cantu (1992). Paternal uniparental disomy of chromosome 15 in a child with Angelman syndrome. *Annals of Neurology,* 32, 512–518.

Nwokah, E. E., H.-C. Hsu, P. Davies, and A. Fogel (1999). The integration of laughter and speech in vocal communication: a dynamic systems perspective. *Journal of Speech, Language, and Hearing Research,* 42, 880–894.

Olsen, H. A. (1976). The use of humor in psychotherapy. *Individual Therapist,* 13, 34–37.

Ostling, R. H. (1994). Laughing for the Lord. *Time,* 15 August 1994, 38.

Overeem, S., G. J. Lammers, and J. G. van Dijk (1999). Weak with laughter. *Lancet,* 354, 838.

Panksepp, J. (1998). *Affective Neuroscience.* Oxford: Oxford University Press.

Panksepp, J., and J. Burgdorf (1999). Laughing rats? Playful tickling arouses high-frequency ultrasonic chirping in young rodents. In S. Hameroff, D. Chalmers, and A. Kaziak (eds.), *Toward a Science of Consciousness III.* Cambridge, MA: MIT Press.

Patterson, P. (1978). The gestures of a gorilla. *Brain and Language,* 5, 72–97.

Pausanias (1971). *Guide to Greece, Vol. 1: Central Greece* (P. Levi, trans.). Baltimore: Penguin.

Pennebaker, J. W. (1980). Perceptual and environmental determinants of coughing. *Basic and Applied Social Psychology,* 1, 83–91.

Pepperberg, I. M. (1987). Acquisition of the same/different concept by an African gray parrot *Psittacus erithacus. Animal Learning and Behavior,* 15, 423–432.

Pierce, R. A. (1985). Use and abuse of laughter in psychotherapy. *Psychotherapy in Private Practice,* 3, 67–73.

Plato (1975). *Philebus* (J. C. B. Gosling, trans.). Oxford: Clarendon Press.

——— (1941). *Republic* (B. Jowett, trans.). New York: The Modern Library.

Plooij, F. (1979). How wild chimpanzee babies trigger the onset of mother-infant play—and what the mother makes of it. In M. Bullowa (ed.), *Before Speech: The Beginning of Interpersonal Communications* (pp. 223–243). Cambridge: Cambridge University Press.

Poeck, K. (1969). Pathology of emotional disorders associated with brain damage. In P. J. Vinken and G. W. Bruyn (eds.), *Handbook of Clinical Neurology,* vol. 3 (pp. 343–367). Amsterdam: North Holland Publishing Company.

Porterfield, A. L. (1987). Does sense of humor moderate the impact of life stress on psychological and physical well-being? *Journal of Research in Personality,* 21, 306–317.

Premack, D. (1972). Language in chimpanzee? *Science,* 172, 808–822.

Preston, G. (1994). The Toronto wave. *Christian Century,* 16 November 1994, 1068–1069.

Provine, R. R. (1973). Neurophysiological aspects of behavior development in the chick embryo. In G. Gottlieb (ed.), *Behavioral Embryology* (pp. 77–102). New York: Academic Press.

——— (1984). Wing-flapping during development and evolution. *American Scientist,* 72, 448–455.

——— (1986). Yawning as a stereotyped action pattern and releasing stimulus. *Ethology,* 72, 109–122.

——— (1989a). Contagious yawning and infant imitation. *Bulletin of the Psychonomic Society,* 27, 125–126.

——— (1989b). Faces as releasers of contagious yawning: An approach to face detection using normal human subjects. *Bulletin of the Psychonomic Society,* 27, 211–214.

——— (1992). Contagious laughter: Laughter is a sufficient stimulus for laughs and smiles. *Bulletin of the Psychonomic Society,* 30, 1–4.

——— (1993). Laughter punctuates speech: Linguistic, social and gender contexts of laughter. *Ethology,* 95, 291–298.

——— (1996a). Contagious yawning and laughter: Significance for sensory feature detection, motor pattern generation, imitation, and the evolution of social behavior. In C. M. Heyes and B. G. Galef (eds.), *Social Learning in Animals: The Roots of Culture* (pp. 179–208). New York: Academic Press.

——— (1996b). Laughter. *American Scientist,* 84, 38–45.

——— (1997a). Bipedalism and speech evolution. *Society for Neuroscience Abstracts,* 23, part 2, 2134.

——— (1997b). Yawns, laughs, smiles, tickles, and talking: Naturalistic and laboratory studies of facial action and social communication. In J. A. Russell and J. M. Fernandez-Dols (eds.), *The Psychology of Facial Expression* (pp. 158–175). Cambridge: Cambridge University Press.

——— (1999). Stand up and talk: Bipedalism and speech evolution. *Society for Neuroscience Abstracts,* 25, part 2, 2170.

Provine, R. R., and K. A. Bard (1994). Laughter in chimpanzees and humans: A comparison. *Society for Neuroscience Abstracts,* 20, part 1, 367.

——— (1995). Why chimps can't talk: The laugh probe. *Society for Neuroscience Abstracts,* 21, part 1, 456.

Provine, R. R., and K. R. Fischer (1989). Laughing, smiling, and talking: Relation to sleeping and social context in humans. *Ethology,* 83, 295–305.

Provine, R. R., and J. A. Westerman (1979). Crossing the midline: limits of early eye-hand behavior. *Child Development,* 50, 437–441.

Provine, R. R., and Y. L. Yong (1991). Laughter: A stereotyped human vocalization. *Ethology,* 89, 115–124.

Randel, D. M. (ed.) (1986). *The New Harvard Dictionary of Music.* Cambridge, MA: Harvard University Press.

Rankin, A. M., and P. J. Philip (1963). An epidemic of laughing in the Bukoba district of Tanganyika. *The Central African Journal of Medicine,* 9, 167–170.

Raper, H. R. (1945). *Man Against Pain.* New York: Prentice-Hall.

Raulin, J.-M. (1900). *Le rire et les exhilarants.* Paris: Libraire J.-B. Bailliere et fils.

Rhodes, R. (1997). *Deadly Feasts.* New York: Simon & Schuster.

Roach, M. (1996). Can you laugh your stress away? *Health,* September 1996.

Rodier, J. (1955). Manganese poisoning in Moroccan miners. *British Journal of Industrial Medicine,* 12, 21–35.

Rohan, J. (1935). *Yankee Arms Maker: The Incredible Career of Samuel Colt.* New York: Harper & Brothers.

Rosen, G. (1968). *Madness in Society.* Chicago: University of Chicago Press.

Rosselli, J. (1984). *The Opera Industry in Italy from Cimarosa to Verdi: The Role of the Impresario.* Cambridge: Cambridge University Press.

Rothbart, M. K. (1976) Incongruity, problem-solving and laughter. In A. J. Chapman and H. C. Foot (eds.), *Humour and Laughter: Theory, Research and Applications* (pp. 37–54). Chichester, England: Wiley.

Rotton, L. (1992). Trait humor and longevity: Do comics have the last laugh? *Health Psychology,* 11, 262–266.

Rotton, J., and M. Shats (1996). Effects of state humor, expectancies, and choice on postsurgical mood and self-medication: A field experiment. *Journal of Applied Social Psychology,* 26, 1775–1794.

Rumbaugh, D. M. (ed.) (1977). *Language Learning by a Chimpanzee: The Lana Project.* New York: Academic Press.

Russell, R. (1992). *Sardonicus.* In C. Baldick (ed.), *The Oxford Book of Gothic Tales* (pp. 435–465). Oxford: Oxford University Press.

Sacks, O. (1987). *The Man Who Mistook His Wife for Hat.* New York: Harper & Row.

Sade, Marquis de (1797/1988). *Juliette* (A. Wainhouse, trans.). New York: Grove Press, p. 797.

Salameh, W. A. (1983). Humor in psychotherapy: Past outlooks, present status, and future frontiers. In P. E. McGhee and J. H. Goldstein (eds.), *Handbook of Humor Research: Vol. 2, Applied Studies* (pp. 61–88). New York: Springer-Verlag.

Sanders, B. (1995). *Sudden Glory*. Boston: Beacon Press.

de Santillana, G. (1961). *The Origins of Scientific Thought.* New York: New American Library.

Savage-Rumbaugh, E. S., and R. Levin (1994). *Kanzi: The Ape at the Brink of the Human Mind.* New York: John Wiley.

Schachter, S., and L. Wheeler (1962). Epinephrine, chlorpromazine, and amusement. *Journal of Abnormal and Social Psychology,* 65, 121–128.

Scheff, T. J., and D. D. Bushnell (1984). A theory of catharsis, *Journal of Research in Personality,* 18, 238–264.

Scheff, T. J., and S. C. Scheele (1980). Humor and tension: The effects of comedy on audiences. In P. M. Tannenbaum (ed.), *The Entertainment Functions of Television.* Syracuse, NY: Erlbaum.

Schopenhauer, A. (1819/1907). *The World as Will and Idea* (R. B. Haldane and J. Kemp, trans.). London: Kegan Paul, Trench, Trabner, & Company.

Selye, H. (1956). *The Stress of Life.* New York: McGraw-Hill.

Shammi, P., and D. T. Stuss (1999). Humor appreciation: A role of the right frontal lobe. *Brain,* 122, 657–666.

Shultz, T. R. (1976). A cognitive-developmental analysis of humor. In A. J. Chapman and H. C. Foot (eds.), *Humour and Laughter: Theory, Research, and Applications* (pp. 12–36). London: J. Wiley.

Siegel, B. S. (1986). *Love, Medicine and Miracles.* New York: Harper & Row.

Siegel, R. K., and A. E. Hirschman (1985). Hashish and laughter: Historical notes and translations of early French investigators. *Journal of Psychoactive Drugs,* 17, 87–91.

Siegman, A. W., and S. C. Snow (1997). The outward expression of anger, the inward experience of anger and CVR: The role of vocal expression. *Journal of Behavioral Medicine,* 20, 29–45.

Simner, M. L. (1971). Newborn's response to the cry of another infant. *Developmental Psychology,* 5, 136–150.

Skinner, B. F. (1986). The evolution of verbal behavior. *Journal of the Experimental Analysis of Behavior,* 45, 115–122.

Spemann, H. (1938). *Embryonic Development and Induction.* New Haven: Yale University Press.

Sroufe, L. A., and E. Waters (1976). The ontogenesis of smiling and laughter: A perspective on the organization of development in infancy. *Psychological Review,* 83, 173–189.

Sroufe, L. A., and J. P. Wunsch (1972). The development of laughter in the first year of life. *Child Development,* 43, 1326–1344.

Starkstein, S. E., R. Migliorelli, A. Teson, G. Petracca, E. Chemerinsky, F. Manes, and R. Leiguarda (1995). Prevalence and clinical correlates of pathological affective display in Alzheimer's disease. *Journal of Neurology, Neurosurgery, and Psychiatry,* 59, 55–60.

Stearns, F. R. (1972). *Laughing.* Springfield, IL: Charles C. Thomas.

Stone, A. A., J. M. Neale, D. S. Cox, A. Napoli, H. Valdimarsdottir, and E. Kennedy-Moore (1994). Daily events are associated with a secretory immune response to an oral antigen in men. *Health Psychology,* 13, 440–446.

Sully, J. (1902). *An Essay on Laughter.* New York: Longmans, Green.

Sweet, W. H. (1959). Pain. In J. Field, H. W. Magoun, and V. E. Hall (eds.), *Handbook of Physiology: Neurophysiology, Vol. I* (pp. 459–506). Washington, DC: American Physiological Society.

Synan, V. (1971). *The Holiness Pentecostal Movement in the United States.* Grand Rapids, MI: William B. Eerdmans.

Tannen, D. (1990). *You Just Don't Understand.* New York: William Morrow.

Tart, C. T. (1971). *On Being Stoned: A Psychological Study of Marijuana Intoxication.* Palo Alto, CA: Science and Behavior Books.

Temerlin, M. K. (1975). *Lucy: Growing Up Human.* Palo Alto, CA: Science and Behavior Books.

Tilney, F., and J. F. Morrison (1912). Pseudobulbar palsy, clinically and pathologically considered, with the clinical report of five cases. *Journal of Nervous and Mental Disease,* 39, 505–535.

Trice, A. D., and J. Price-Greathouse (1986). Joking under the drill: A validity study of the coping humor scale. *Journal of Social Behavior and Personality,* 1, 265–266.

Turnbull, C. M. (1961). *The Forest People.* New York: Simon & Schuster.

Van Lawick-Goodall, J. (1968). Behaviour of free-living chimpanzees of the Gombe Stream area. *Animal Behaviour Monographs,* 1, 163–311.

Villard, F. (1872). *Du hachisch.* Paris: Adrien Delahaye.

Warner, E. (1985). *Heroes, Monsters and Other Worlds from Russian Mythology.* London: Peter Lowe.

Washburn, R. W. (1929). A study of the smiling and laughing of infants in the first year of life. *Genetic Psychology Monographs,* 6, 397–535.

Weaver, J. B., J. L. Masland, S. Kharazmi, and D. Zillmann (1985). Effect of alcoholic intoxication on the effects of different types of humor. *Journal of Personality and Social Psychology,* 49, 781–787.

Wechsberg, J. (1945). *Looking for a Bluebird.* Boston: Houghton Mifflin.

Weisberger, B. A. (1958). *They Gathered at the River.* Boston: Little, Brown.

Weiskrantz, L., J. Elliot, and C. Darlington (1971). Preliminary observations on tickling oneself. *Nature,* 230, 598–599.

White, S., and P. Camarena (1989). Laughter as a stress reducer in small groups. *Humor,* 2, 73–79.

White, S., and A. Winzelberg (1992). Laughter and stress. *Humor,* 5, 343–355.

Williams, C. A., and J. L. Frias (1982). The Angelman ("happy puppet") syndrome. *American Journal of Medical Genetics,* 11, 453–460.

Willis, E. F. (1992). Nitrous oxide: Dentistry's own special addition. In A. G. Williams (ed.), *Denistry Faces Addition.* St. Louis: Mosby Yearbook.

Wilson, C. D. (1931). *A Study of Laughter Situations Among Young Children.* Unpublished doctoral dissertation. Lincoln: University of Nebraska.

Wilson, S. A. K. (1924). Pathological laughing and crying. *Journal of Neurology and Psychopathology,* 4, 299–333.

Wolfe, R. J., and L. F. Menczer (eds.) (1994). *I Awaken to Glory.* Boston: Boston Medical Library.

Yerkes, R. M. (1943). *Chimpanzees, a Laboratory Colony.* New Haven: Yale University Press.

Zigas, V. (1990). *Laughing Death: The Untold Story of Kuru.* Clifton, NJ: Humana Press.

Zillmann, D., and J. Bryant (1980). Misattribution theory of tendentious humor. *Journal of Experimental Psychology*, 16, 146–160.

Zillmann, D., S. Rockwell, K. Schweitzer, and S. Sundar (1993). Does humor facilitate coping with physical discomfort? *Motivation and Emotion,* 17, 1–21.

Ziv, A. (1984). *Personality and Sense of Humor.* New York: Springer Publishing Company.

Zotterman, Y. (1939). Touch, pain and tickling: An electrophysiological investigation on cutaneous nerves. *Journal of Physiology,* 95, 1–28.

INDEX